FOCUS ON BLACK HOLE RESEARCH

FOCUS ON BLACK HOLE RESEARCH

PAUL V. KREITLER
EDITOR

Nova Science Publishers, Inc.
New York

For permission to use material from this book please contact us:
Telephone 631-231-7269; Fax 631-231-8175
Web Site: http://www.novapublishers.com

Library of Congress Cataloging-in-Publication Data
Available upon request

ISBN 1-59454-460-3

Published by Nova Science Publishers, Inc. ✢New York

CONTENTS

PREFACE

A black hole is a point of extreme mass in spacetime with a radius, or event horizon, inside of which all electromagnetic radiation (including light) is trapped by gravity. A black hole is an extremely compact object, collapsed by gravity which has overcome electric and nuclear forces. It is believed that stars appreciably larger than the Sun, once they have exhausted all their nuclear fuel, collapse to form black holes: they are "black" because no light escapes their intense gravity. Material attracted to a black hole, though, gains enormous energy and can radiate part of it before being swallowed up. Some astronomers believe that enormously massive black holes exist in the center of our galaxy and of other galaxies. This new book brings together leading research from through-out the world.

In Chapter 1, *Black Holes or Eternally Collapsing Objects: A Review of 90 Years of Misconceptions*, *Abhas Mitra* critically reviews the concepts which gave rise to the Black Hole (BH) paradigm over the past 90 years. First the author points out that the very attempt to find the spacetime around a "Massenpunkt" (Point Mass) implicitly forced the entry of Point Singularities in General Relativity (GR). Iit is this borrowed Newtonian idea of a "Point Mass" which directly laid the foundation of the BH paradigm in GR. The author shows that in order that neither any geometrical nor any physical scalar related to the metric associated with a Massenpunkt diverges in a region which is not a physical singularity, the mass of an uncharged Massenpunkt (i.e., the BH), obtained as an integration constant, must be $M_0=0$. This result agrees exactly with the result found by Arnowitt, Deser & Misner way back in 1962: "a neutral particle (one coupled only to gravitational field) has a zero clothed mass, showing that the mass of the particle arises entirely from its interaction with other field" (however, for a neutral finite body, of course there would be finite clothed mass). If the point particle is charged, again, it would have finite mass, and it can be shown, that, mathematically GR allows existence of charged BHs. But, astrophysical BHs are uncharged, and, in this review a BH would mean a non-rotating neutral BH unless otherwise mentioned.

Einstein's general theory of relativity poses many problems to the quantum theory of point particle fields. Among them is the fate of a massive point particle. Since its rest mass exists entirely within its Schwarzschild radius, in the classical solutions of Einstein's theory, the respective system should be a black hole. B.F.L. Ward in

Chapter 2, *Black Holes and Massive Elementary Particles in Resummed Quantum Gravity,* addresses this issue using exact results in a new approach to quantum gravity based upon well-tested resummation methods in point particle quantum field theory. The author shows that the classical conclusion is obviated by quantum loop effects. He shows that our new approach already passes two theoretical checks with the published literature; for, it reproduces known results on the one-loop correction to the graviton self-energy in scaler matter coupled to Einstein's gravity as analyzed by 't Hooft and Veltman and it is consistent with the asymptotic safety results of Bonanno and Reuter on the behavior of Newton's constant in the deep Euclidean regime. Indeed, his approach is consistent with the black hole phenomenology of the latter authors, including their results on the final state of the Hawking radiation for an originally massive black hole. Further black hole related phenomenological implications are also discussed.

Motivated by the recent interest in quantization of black hole area spectrum, M.R. Setare considers the area spectrum of Schwarzschild, BTZ, extremal Reissner-Nordström, near extremal Schwarzschild-de Sitter, and Kerr black holes in *Quasinormal Modes, Reduced Phase Space and Area Spectrum of Black Holes*. Based on the proposal by Bekenstein and others that the black hole area spectrum is discrete and equally spaced, Setare implements Kunstatter's method to derive the area spectrum for these black holes. The author shows that although the Schwarzschild black hole spectrum is discrete, it is non equispaced in general. In the other hand the reduced phase space quantization is another technique is discussed here. However there is a discrepancy between the result of the reduced phase space methodology and quasinormal modes approach for area spectrum of some black holes.

In Chapter 4, *What Black Holes Can Teach Us, Sabine Hossenfelder* discuss black holes and how they merge together different fields of physics. From General Relativity over thermodynamics and quantum field theory, they do now also reach into the regime of particle and collider physics. In the presence of additional compactified dimensions, it would be possible to produce tiny black holes at future colliders. The author tests Planck scale physics and the onset of quantum gravity. The understanding of black hole physics is a key knowledge to the phenomenology of these new effects beyond the Standard Model.

A. S. Majumdar, in the last chapter, *Cosmology with Primordial Black Holes*, considers the evolution of primordial black holes formed due to the gravitational collapse of overdense regions in the early universe. These black holes are formed with an initial mass spectrum, and lose energy by the process of Hawking evaporation. However, the process of accretion of radiation in the radiation dominated era could under certain conditions counterbalance the effect of evaporation, and this leads to longer lifetime for the black holes. He investigates the scenarios of the early universe with a population of primordial black holes interacting with the surrounding radiation in standard cosmology and as well as in braneworld cosmology separately. In the braneworld scenario the 5-dimensonal black hole geometry leads to modified evaporation and efficient accretion that is particularly conducive to the survival of primordial black holes up to several cosmologically interesting eras. Primordial black holes in the braneworld scenario could act as candidates for non-baryonic cold dark matter. A mechanism for the formation of binaries with primordial braneworld black holes is also discussed.

In: Focus on Black Hole Research
Editor: Paul V. Kreitler, pp. 1-94

ISBN 1-59454-460-3

Chapter 1

Black Holes or Eternally Collapsing Objects: A Review of 90 Years of Misconceptions

Abhas Mitra*
Nuclear Research Laboratory, Bhabha Atomic Research Center
Mumbai-400085, India

Abstract

We critically review here the concepts which gave rise to the Black Hole (BH) paradigm over almost past 90 years. First we point out that the very attempt to find the spacetime around a "Massenpunkt" (Point Mass) implicitly forced the entry of Point Singularities in General Relativity (GR). And it is this borrowed Newtonian idea of a "Point Mass" which directly laid the foundation of the BH paradigm in GR. We show that in order that neither any geometrical nor any physical scalar related to the metric associated with a Massenpunkt diverges in a region which is not a physical singularity, the mass of an uncharged Massenpunkt (i.e., the BH), obtained as an integration constant, must be $M_0 = 0$. This result agrees exactly with the result found by Arnowitt, Deser & Misner way back in 1962: "a neutral particle (one coupled only to gravitational field) has a zero clothed mass, showing that the mass of the particle arises entirely from its interaction with other field" (however, for a neutral finite body, of course there would be finite clothed mass). If the point particle is charged, again, it would have finite mass, and it can be shown, that, mathematically GR allows existence of charged BHs. But, astrophysical BHs are uncharged, and, in this review a BH would mean a non-rotating neutral BH unless otherwise mentioned.

Simultaneously we show that even for the spherical gravitational collapse, no "trapped surface" is ever formed for collapse of isolated bodies and thus all the "Singularity Theorems" get invalidated for isolated bodies. Physically, trapped surfaces cannot occur in order that the local 3-speed of the collapsing matter (w.r.t. a non-comoving observer) $v < c$, the speed of light. Thus collapse process of massive

*E-mail address: amitra@apsara.barc.ernet.in, abhasmitra@rediffmail.com

bodies can at most asymptotically (infinite comoving proper time) approach the state of a zero mass BH. At a given epoch, the collapsing body would be "Hot" and of finite mass. The pressure of the gravitationally trapped radiation within the body grows as $\sim (1+z)^2$. Consequently catastrophic collapse gets dynamically halted as the luminosity of trapped radiation attains its local Eddington value. We call such objects as Eternally Collapsing Objects (ECOs) and they are in quasistatic equilibrium due to pressure of trapped radiation. In addition, if the magnetic field energy density could be comparable to the radiation pressure it may also contribute to the hydrostatic balance. The conventional mass upper limit of COLD and strictly STATIC objects is simply not applicable to ECOs because they are HOT and not strictly static. In a strict sense, the external spacetime of ECOs is given by radiating Vaidya Metric instead of static radiationless Schwarzschild (actually Hilbert) metric. Hence the Buchadal limit of $z < 2.0$, where z is the surface gravitational redshift of the compact object, is not valid for the ECOs. On the other hand, they can have arbitrary high value of z, which in the stellar mass cases could be $z \sim 10^{7-8}$.

We point out that Quantum Mechanics (QM) demands that there is an upper limit of the Proper Acceleration, $a_p = c^2/2l_p$, where l_p is the Planck length. Thus if QM is invoked, in order that the acceleration of a test particle is bounded by a_p, there cannot be any EH or any BH at all. Consequently, if QM is applied to GR, actually, there is no Hawking Radiation and no Quantum Information Loss Paradox at all (classically also there is no Hawking Radiation because there is no finite mass BH). Therefore as per Quantum Gravity, gravitational collapse at the most result in Ultra Compact Objects with radius $R_{QM} \to R_g$ in accordance with the idea of ECOs and the minimum value of $R_{QM} \sim l_p$. We also mention that the so-called rotating (Kerr) BHs too have M=0 and in fact they cannot rotate at all. Therefore, the spinning massive objects found by astronomers must be rotating ECOs/MECOs. Finally, by appealing to the basic principle of *Invariance of 4-Volume* in any curvilinear coordinate transformations, in a surprisingly simple manner, we directly show that, the *integration constant* $\alpha_0 = 2M_0$ which appeared in the vacuum Hilbert/Schwarzschild solution of the spacetime for a Massenpunkt 90 years ago is **actually zero**. Since the entire BH paradigm has been built on the Newtonian assumption of $\alpha_0 > 0$, the paradigm, thus, collapses for ever with this simple direct result. Hence there is no massenpunkt, no BH, no singularity (for isolated collapse) in GR. We also emphasize that the "Dark Energy Stars", proposed by various authors as an alternative to BHs are unphysical. And the only alternative to BHs are ECOs and their Ultra Magnetized versions MECOs. We also point out why collapse of sufficiently massive objects must generate EOCs or their ultramagnetized version MECOs rather than anything else.

1 Introduction

"If you are out to describe the truth, leave elegance to the tailor"
- Albert Einstein

Ever since Newton discovered gravitation as an universal and, further, a non-linear phenomenon, the fate of sufficiently massive isolated bodies pulled inwards by self-gravity became a topic of scientific concern. Though the broader scientific community is still grappling with an answer to question[1] the present review would highlight that there is already a definitive answer to this question - the gravitational collapse would either form Ultra Compact Objects with physical surfaces in a finite comoving proper time or else the collapse would proceed indefinitely as the spacetime membrane would be stretched indefinitely by infinitely strong spacetime curvature[2,3,4,5]. In short, this review will discuss Gravitational Collapse (GC) and related concepts like those of so-called Black Holes (BH), Naked Singularity (NS), Ultra Compact Objects(UCO) and Eternally Collapsing Objects (ECO).

In this review we would unfurl the developments of this old subject in an approximately chronological mode. In order that any massive body stays put against its self-gravity, the magnitude of the negative self-gravitational energy (E_g) must be either negligible or tentatively balanced by internal energies. While all internal energies are intensive thermodynamical quantities, i.e., if two parts of the body have internal energies U_1 and U_2, the combined internal energy will be $U = U_1 + U_2$, the self-gravitation energy grows rapidly in a non-linear manner and hence massive bodies are unstable to GC.

For planets or less massive bodies with $M \ll 0.01 M_\odot$ (where $M_\odot$ stands for solar mass) and density $\rho \sim few\ gm\ cm^{-3}$, usually, $|E_g| \ll U$, which is mostly of chemical origin. And obviously there is no question of any generation of internal energy by nuclear means for such low mass objects. For more massive objects such as Brown Dwarfs ($\sim 0.01 M_\odot$) again there is no nuclear energy generation. But in this mass range, $|E_g|$ tends to be of same order of magnitude as U and such objects tend to contract on a very long secular period. The same is true for low density giant massive clouds which are the progenitors of all stars. There is no nuclear or chemical energy generation in the clouds. It is extremely important to understand that given sufficiently low density, there could be self-gravitating configurations of arbitrary high mass and which can remain in quasistatic hydrostatic equilibrium for millions or even billions of years even in the absence of any nuclear energy generation at their cores.

Then how do they manage to survive again catastrophic GR? Although all astrophysicists learn how this is possible, it seems that this lesson is forgotten by many General Relativists while they study the phenomenon of GC and UCOs.

1.1 Helmholtz -Kelvin Process

In the simplest case, if we consider a quasistatic monoatomic self-gravitating object (i.e., no internal degree of freedom and having an effective ratio of the specific heats $\gamma = 5/3$), it follows from the Virial Theorem[6] that

$$2U + E_g = 0; \qquad U = -\frac{1}{2} E_g \tag{1}$$

Now as the configuration contracts secularly, the value of $|E_g|$ increases. If the increase in $|E_g|$ is $|\Delta E_g|$, the corresponding change in internal energy will be

$$\Delta U = -\frac{1}{2} \Delta E_g = +\frac{1}{2} |\Delta E_g| > 0 \tag{2}$$

Thus *even when there is no chemical or nuclear or any other fuel present, gravitational contraction generates fresh internal energy and attendant pressure gradient*, i.e., gravitational contraction generates its own antidote, namely, additional pressure gradient which opposes catastrophic GR. Unless one understands, appreciates as well as applies this magical property of self-gravity, all studies of GC are liable to be completely incorrect and misleading.

The total energy of the isolated body

$$E = U + E_g \tag{3}$$

accordingly changes by

$$\Delta E = \Delta U + \Delta E_g = +\frac{1}{2}|\Delta E_g| - |\Delta E_g| = -\frac{1}{2}|\Delta E_g| < 0 \tag{4}$$

Thus the total energy of the isolated body decreases and hence in order to honor total energy conservation, the contracting body **must radiate** an amount of energy $(1/2)|\Delta E_g|$ during this process.

For an arbitrary value of γ, there will be following modifications[6]:

$$\Delta U = +\frac{1}{3(\gamma - 1)}|\Delta E_g| > 0 \tag{5}$$

and

$$\Delta E = -\frac{3\gamma - 4}{3(\gamma - 1)}|\Delta E_g| < 0 \tag{6}$$

The H-K process will cease to exist only if one would strictly attain a state of $\gamma = 4/3$. But in reality, this limit is never strictly attained because only for a pure incoherent photon gas $\gamma = 4/3$.

Thus when we deal with physical matter having pressure and internal energy,

(i) GR itself generates its own antidote, i.e., fresh source of internal energy even if there would be no source of fresh generation of nuclear or chemical energy

and

(ii) There cannot be any GC without accompanying emission of radiation. Conversely, for matter with pressure and internal energy, the assumption of an *adiabatic collapse is absolutely unphysical*.

Such an assumption tantamounts to the denial of existence of internal energy and pressure, i.e., adiabatic collapse is possible only for a pressureless case! Thus when, in the mathematical equations of GC, one considers a finite p, *there will not be any collapse at all if radiative flux q would be set to zero*. In section 13, we will prove this statement using exact GR collapse equation.

For bodies with $\rho \sim$ few g cm^{-3}, as the mass increases to $\sim 0.1 M_\odot$, H-K process heats up the core so much that nuclear fuel gets ignited and a main-sequence stellar phase sets in. At this stage, burning of the nuclear fuel may keep the main-sequence star in strict hydrostatic equilibrium and the H-K process might get a breather. However whenever the nuclear burning would be switched off for whatever reason, the H-K process must be activated. It is because of this reason that the duration of the neutrino burst for SN 1987A was around 10 s when the corresponding free fall time scale of the stellar iron core would be $<$ 1 ms.

In this review we will go on describing various crucial developments which consolidated the present paradigm of BHs and GC over the past 90 years and try to analyze them by unmasking the overt or covert assumptions and simplifications in a more or less chronological manner. At several stages, and from several independent *exact* derivations, it would be seen that, though, formally, the idea of a BH seems to be natural, for overall physical and mathematical consistency, the (uncharged) BHs must have the unique gravitational mass $M = 0$! Howsoever one would feign to ignore it at the present moment there is no escape from this result. Since comoving proper time of formation of such zero mass BHs is infinite, in practice, even these zero mass BHs are never formed at all. Further since all the astronomical and other objects (even most of the elementary particles) have finite gravitational masses, none of them are BHs. Accordingly, the so-called BH Candidates (BHCs) whose masses are well above the upper mass limit of **cold** and **strictly static** compact objects self-gravitating objects, too, cannot be true BHs even though they could be more compact than conventional **cold** and static Neutron Stars.

2 Pre-Relativistic "Dark Stars"

In the Newtonian case, the collapse of a self-gravitating spherical fluid, by *ignoring radiation pressure and radiative transport*, is determined by these following equations:

$$\frac{\partial \rho}{\partial t} + \nabla.(\rho \vec{v}) = 0 \tag{7}$$

$$\rho \frac{d\vec{v}}{dt} = \vec{f} - \nabla p \tag{8}$$

and

$$\nabla^2 \phi = 4\pi G \rho \tag{9}$$

where ϕ is the Newtonian potential and $\vec{f}$ is the appropriate "body force" due to self-gravity. In the Newtonian formalism, the H-K process cannot be meaningfully incorporated, and, tentatively, one might write $p = p_g + p_r$ where p_g is the gas pressure and p_r is the radiation pressure. In the approximation of radial propagation of heat flux, one might tentatively write

$$p_r = q = \frac{L}{4\pi R^2} \tag{10}$$

where L is the luminosity. Even by accepting $p_r = 0$, i.e., the absence of any H-K mechanism, contrary to the popular misconception, there cannot be any exact solution to these equations because neither the *evolving* Equation of State (EOS) nor the density profile of the collapsing fluid is *exactly* known. Obviously the opacity and radiation transport properties are not known either. An "exact" solution is possible only if one would assume the fluid has (i) no pressure, $p \equiv 0$, (ii) there is no heat transport at all ($q = 0$) and (iii) there is no H-K mechanism at all.

One might pretend here a somewhat more general treatment by assuming (i) a polytropic EOS $p = K\rho^{1+1/n}$ with a fixed value of polytropic index $n = n$ and (ii) an adiabatic collapse with no heat transport ($q = 0$). Since n is bound to change, and EOS cannot always be represented in this simple polytropic fashion, this approach cannot be called as an *exact one*. Moreover as discussed in the previous section, the assumption of adiabaticity tantamounts to denial of existence of pressure and internal energy. Hence actually the *mathematical* treatment of an *adiabatic GC, physically, implies, no GC at all!*

In any case, when one neglects pressure completely, one would have *free fall* and the fluid would collapse to a geometrical point in a time:

$$T_c = \frac{\pi}{2}\left(\frac{3}{8G\pi\rho(0)}\right) \tag{11}$$

where $\rho(0)$ is the central density at $T = 0$ when the cloud is assumed to be at rest.

It is often mentioned in the text books that once Sun would exhaust its nuclear fuel, it would undergo approximate free fall on a time scale governed by the above formula, i.e., Sun would collapse within a few minutes. In reality, when one would invoke physics and not treat GC as an arena for convenient workable applied mathematics exercises, even without any GR concept such as stretching of spacetime membrane by increasing gravitation, it is known that the Sun would keep on shining for a H.K. time scale of[6]

$$t_{H.K} = b\frac{3\gamma - 4}{3(\gamma - 1)}\frac{GM^2}{LR} \tag{12}$$

where b is a numerical constant of the order of unity; and if the configuration is approximated as a polytrope of index n, then

$$b = \frac{3}{5 - n} \tag{13}$$

How does it happen? This happens because as the Sun would try to contract, the magical property of its self-gravitation would generate fresh pressure gradient within it and would retard the collapse process. At the same time, it *must* radiate and otherwise there would not be any collapse at all! If it is assumed that the self-gravitation generated luminosity would remain fixed at the present nuclear generated luminosity of $L = L_\odot = 2 \times 10^{33}\ erg/s$, and further, the radius of Sun would remain fixed at $R = R_\odot$, the present energy reserve would last for the H-K time scale of

$$t_{H.K.} \sim 2.4 \times 10^7\ year \sim 1.5 \times 10^{13}\ min \tag{14}$$

Note that even without any finer details, the result obtained by free-fall assumption is **less by an approximate factor of** 10^{12}. This simple example shows that all the conventional conclusions about GC obtained by using various covert simplifying assumptions could be completely incorrect.

Eq.(14) implicitly assumes that (i) Sun would continue to shine with its present luminosity, (ii) its radius R would remain more or less same (iii) Newtonian expression for gravitational potential energy would remain valid and (iv) there would not be any stretching of spacetime (a GR effect).

When all such assumptions are removed, particularly since R would keep on decreasing, actual value of $t_{H.K.}$ may dramatically increase. One can wonder, whether this non-linear H-K. process could be a runaway one, without terminating ever? Of course, there cannot be any answer to this tempting question in the Newtonian framework.

However, if we ignore such subtle points, since M remains fixed in the Newtonian gravity (because of the absence of $E = Mc^2$ law), and R keeps on decreasing, it is legitimate to conceive of a situation when one would have

$$R = R_g = \frac{2GM}{c^2} \tag{15}$$

when the escape speed from the surface of the collapsing object would be equal to the speed of supposed light corpuscles. Following this stage, light will not be able to escape the configuration and one would have a "Dark Star", and as is well known, this idea was originally due to John Michell[7].

Note that, even in GTR, if one would assume an adiabatic collapse, M would remain fixed despite the $E = Mc^2$ law and one would obtain the same **Newtonian Eq.(15)**. And it is such an explicit or implicit assumption of adiabaticity (by treating q as a sufficiently small quantity) which has put GC research on a completely misleading track over the past 90 years.

3 Hilbert and Schwarzschild Solutions

John Michell's idea of a Dark Star was largely ignored until Hilbert and Schwarzschild found solutions (using different gauges) for the gravitational field around a "Massenpunkt", i.e., a point mass. But could there be "Point Mass" in GR in the first place? Even if the idea of a "Point Mass" (of finite gravitational mass M_0) would be admissible in GR, there is no denying the fact that this idea of a "Massenpunkt" was carried forward from Newtonian physics (remember, in String Theories, there is no "Point Mass" at all). On the other hand, even if the idea of a "Point Mass" would appear to be untenable in GR, because of Birchoff's Theorem, the gravitational field around a non-radiating spherically symmetrical body may be obtained by considering the existence of an equivalent "Point Mass" at its center of symmetry. In fact GR permits existence of "Point Masses" incase they would be charged.

3.1 Spherically Symmetric Vacuum Spacetime

In GTR, because of its intended generality to handle arbitrary (classical) physical problems involving arbitrary observers and accelerations, the choice of a coordinate system is completely arbitrary. And in principle it is possible to use transformations which would mix up even t and supposedly angular coordinates θ, ϕ. Thus, in a strict sense, coordinates, in GTR, are mere labels and need not have any physical or geometrical meaning. However such general coordinates (labels) may not "adapt" the geometrical symmetry of the problem. Further, it should be borne in mind, that, in the context of a given physical problem, we do treat that "given physical problem" and need not pretend an intellectual/mathematical exercise where the coordinates must be arbitrary labels. Since GTR is physics and not just applied mathematics and has been developed to seek correct answer to physical problems, it would be desirable to have coordinates directly related to observations or physical quantities. At least, in case, one is indeed using coordinates which have physical and geometrical significance apart from being mere "labels", there is no reason, why should such coordinates would have to be discarded in favor of coordinates devoid of any direct physical or geometrical significance.

For further appreciation of this topic, recall that the most general form of metric in spherical symmetry which recognizes the conventional polar coordinate θ and azimuth ϕ[8,9]:

$$ds^2 = h(r,t)dr^2 + k(r,t)(\sin^2\theta d\phi^2 + d\theta^2) + l(r,t)dt^2 + a(r,t)drdt \qquad (16)$$

But r and t are still arbitrary radial and time coordinates, i.e., mere labels. h, k and a are arbitrary functions of the indicated arguments. Because of the arbitrariness in the choice of reference system in GR, we can subject these coordinates to any transformation which *does not* destroy the central symmetry of the metric; hence it is possible to conceive of a transformation

$$r = f_1(r', t'); \qquad t = f_2(r', t') \qquad (17)$$

and eliminate the $drdt$ term in Eq.(16). The coordinate t can still be subject to an arbitrary transformation $t = f_3(t')$ not containing r. Further let us write

$$k(r,t) = -R^2(r,t) \qquad (18)$$

The above condition implies that R is defined in such a way that the circumference of a circle with its center at the origin of coordinates is $C = 2\pi R$. Also, $S = 4\pi R^2$ is the INVARIANT area of 2-spheres around the center of symmetry:

$$R = C/2\pi = \sqrt{S/4\pi} \qquad (19)$$

Hence if a light pulse would emanate around the center of symmetry, its intensity would decrease as $\propto R^{-2}$. Thus R has both physical and geometrical significance and is not just a "label". This latter property of R is exactly the same as the Newtonian radial coordinate has and thus it is expected that the radial coordinate appearing in the GR version of Kepler's 3rd law, would directly contain R rather than anything else. Such geometrical and physical properties of R, encapsulated in Eqs.(18-19), are valid *everywhere* irrespective of whether the region is "trapped" or not. This strongly suggests that R must remain a spacelike coordinate everywhere and if in a certain mathematical scenario, R would appear to lose such a property, that mathematical scenario must not be realized in physical universe.

Also writing (henceforth we take $c = 1, G = 1$ unless mentioned otherwise)

$$h(r,t) = e^{\nu}, \qquad l(r,t) = -e^{\lambda} \tag{20}$$

we have

$$ds^2 = e^{\nu}dt^2 - e^{\lambda}dr^2 - R^2(d\theta^2 + \sin^2\theta d\phi^2) \tag{21}$$

The solution for the gravitational field associated with a "Massenpunkt" as found by Hilbert is[10]

$$R = R(r) \tag{22}$$

$$e^{\nu(r,t)} = 1 - \frac{\alpha_0}{R} \tag{23}$$

$$e^{\lambda(r,t)} = \left(1 - \frac{\alpha_0}{R}\right)^{-1} \tag{24}$$

where the **integration constant** $\alpha_0 = 2M_0$, and M_0 is the gravitational mass of the massenpunkt. For an extended body (not a point mass), either by means of Kepler's law or by reducing the field equations to the corresponding Newtonian equations in the weak field limit, one obtains $\alpha_0 \to \alpha = 2M$ where M is the gravitational mass of the extended body. For a Massenpunkt too one follows the same procedure (though even in the Newtonian case, there is no uncharged point mass in reality) and interprets $\alpha_0 = 2M_0$, where M_0 is the gravitational mass of the massenpunkt. Recall that for an extended static spherical body of external radius R_0, we have

$$M = \int_0^{R_0} 4\pi R^2 \rho dR \tag{25}$$

and this is of course finite. However there is no certainty that

$$M_0 = \lim_{R_0 \to 0} M \tag{26}$$

is finite and it is quite possible that M_0 is actually zero. Note here that, since any self gravitating body must have a pressure gradient, in GTR, in a strict sense, there must be a density gradient too, because otherwise the sound speed $c_s = \sqrt{dp/d\rho} \to \infty$. And, in a strict sense, the only fluid which can have uniform density ρ_0 ($d\rho = 0$) is a dust for which $p = dp/dR = 0$. This is so because, if the "dust" is really considered as a "fluid", in order that $c_s \neq 0$, one must have $d\rho/dR = 0$ too. Therefore, *if a dust is considered as a fluid, it must be considered as homogeneous too*. For all other (physical) fluids, there cannot be any homogeneous configuration in GR. Further, in order that, $c_s \neq 0$ at the boundary of the dust ball (considered as a fluid) where $dp = 0$, one must again have $d\rho = 0$. But since $\rho = 0$ just outside the dust ball (or any non-radiating isolated body), one must have $\rho = 0$ within the dust ball too. So, we explicitly see that $M = 0$ for a dust ball (if it would be considered as a fluid).

The extremely important solution (22-24) due to Hilbert, however, got ascribed to the name of Schwarzschild and even when it was recognized that this solution is actually due to Hilbert, he received only brickbats rather than any bouquets[12,13,14]!

For all spherically symmetric geometry, the Invariant Circumference around the point of symmetry is $2\pi R$. Accordingly, in the Hilbert solution, the radial coordinate is R, i.e., the circumference coordinate, and the center of symmetry lies at $R = 0$, the site of the massenpunkt. Therefore it appears that Hilbert's choice of gauge and associated radial coordinate is in perfect harmony with the spherical symmetry.

Schwarzschild's solution for the Massenpunkt formally looks exactly same as Eqs.(22-24). However his choice of gauge was different and in his view the radial coordinate r is related to Invariant Circumference coordinate R through the following relation:

$$R^3 = r^3 + \alpha_0^3; \qquad r^3 = R^3 - \alpha_0^3 \tag{27}$$

Here the massenpunkt is sitting at $r = 0$ i.e., at $R = \alpha_0$. The Invariant Circumference just skirting the center of coordinate system ($r = 0$), $C_0 \neq 0$, but $C_0 = 2\pi\alpha_0$! This would mean that an external observer would perceive the radius of the point mass as not zero but α_0! Thus it seems that Schwarzschild's choice of gauge could be acceptable, for a point mass, only when $\alpha_0 = 0$! And for an extended body, since $\alpha \neq 0$, there would be no chance of accepting the Schwarzschild gauge.

Further, it is only when R is interpreted as the radial coordinate, one *recovers Kepler's law (for an extended body) in an exact fashion*:

$$\omega^2 = \frac{\alpha}{2R^3} \propto R^{-3} \tag{28}$$

where $\omega = d\phi/dT$ is the angular speed. On the other hand, if one would interpret the Schwarzschild radial variable as the radial coordinate, because of its off-center character, one *does not obtain the exact Kepler's law*:

$$\omega^2 = \frac{\alpha}{2(r^3 + \alpha^3)} \tag{29}$$

i.e., ω is not $\propto r^{-3/2}$. Here we have replaced $\alpha_0 \to \alpha$ in order to indicate that Sun or other extended bodies are not massenpunkts though by virtue of Birchoff's theorem[15] the external gravitational field may be considered to be due to massenpunkt with mass M. Despite these anomalies associated with the original Schwarzschild solution, some authors strongly advocate that Hilbert's solution (which is in harmony with the central symmetry) should be discarded and Schwarzschild's solution should be considered as the valid solution[12-14]. These authors argue so because of the following main reasons:

- As can be seen from Eqs.(21-24), the metric has singularities both at $R = 0$ and at $R = \alpha_0$. As per Hilbert solution, $R = 0$ singularity is at the location of the "Massenpunkt", something like the case in Newtonian physics. But the singularity at $R = \alpha_0$ is an additional one (unless $\alpha_0 \equiv 0$) and the notion of BHs has been based on this solution which considers $R = \alpha_0$ surface to be the Event Horizon (EH) of the BH. The corresponding singularity at $R = \alpha_0$ is considered, in this paradigm, to be a mere "Coordinate Singularity".

• However some of these authors[12,13] have correctly shown that this view is incorrect and the singularity at $R = \alpha_0$ is a genuine physical singularity because the Scalar constructed by using 4-acceleration of a test particle and which signifies *a local invariant property of the associated spacetime, diverges* there. Since there is (at the most) one genuine singularity in the problem and which could be only at the location of the "Massenpunkt", these authors consider that the Massenpunkt is at $R = \alpha_0$, which corresponds to $r = 0$. This is the reason that these authors are keen to reinstate the original Schwarzschild solution by discarding the Hilbert solution which, though, is in perfect conformity with central symmetry and yields the correct form of Kepler's law!

In our view though the concern of this authors that the $R = \alpha_0$ singularity is a physical singularity is justified, their suggested remedy, i.e., to accept Schwarzschild solution with its inherent off-center character and associated oddities is incorrect. This conundrum has many nuances, like the validity of the very concept of a "point mass", and it cannot be solved by crudely discarding one gauge and accepting another irrespective of its oddities and inconsistencies. Even if one would accept the original Schwarzschild solution, there will, at the most, be a window dressing of the problem; an inherent physical anomaly can neither be established nor removed by merely resorting to the freedom of choice of gauge.

As we shall discuss shortly, all the geometrical scalars such as the Kretschmann scalar varies as $K \sim R^{-6}$ and not as r^{-6} which again shows that it is the Scalar variable R which is best suited as the radial coordinate rather than r. These authors argue that $R = \alpha_0$ is the true physical singularity; then why doesn't $K \to \infty$ at $R = \alpha_0$ (unless $\alpha_0 \equiv 0$)? These authors have no answer for this physical question.

Similarly why should K blow up at $r = -\alpha_0$ (i.e., at $R = 0$) and not at $r = 0$ when the Massenpunkt is supposed to be sitting at $r = 0$? More fundamentally, if $r = 0$ is really the natural origin of the coordinate system, why should a negative radius ($r = -\alpha_0$) arise at all? There is no remedy for such physical and mathematical oddities in the Schwarzschild gauge.

Further why should the intensity of a spherical pulse of light emitted by a spherical body fall off as R^{-2} and not as r^{-2}? In other words, why should the pulse of light would appear to be propagating in an aspherical manner to an observer sitting at the center of the body? There is no solution for such physical questions in this picture. Also, while these authors, correctly point out the importance of divergence of Acceleration Scalar at the EH, their overall approach to the problem is not very physical because they do not address the real life problem of gravitational collapse of massive stars at all and they reduce the problem to an applied mathematics one of choosing an appropriate gauge.

Thus we reject the original Schwarzschild gauge as unphysical and stick to the Hilbert solution (Massenpunkt at $R = 0$) which is however mistaken as the Schwarzschild solution. It is pleaded that now Hilbert be given his due credit by renaming his solution correctly.

By renaming the temporal coordinate in the Hilbert metric as T, it becomes:

$$ds^2 = \left(1 - \frac{\alpha_0}{R}\right) dT^2 - \left(\frac{dR^2}{1 - \alpha_0/R}\right) - R^2(d\theta^2 + \sin^2\theta d\phi^2) \tag{30}$$

To explore the physical meaning of R and T further, note that, the infinitesimal proper time interval measured by a local $R, \theta, \phi = fixed$ observer is

$$ds = d\tau = \sqrt{1 - \alpha_0/R}\, dT \tag{31}$$

or,

$$dT = \frac{ds}{\sqrt{1 - \alpha_0/R}} = \frac{d\tau}{\sqrt{1 - \alpha_0/R}} \tag{32}$$

Here R and α_0 are *scalars*. Further since proper time interval $d\tau$ is a physically measurable quantity, dT, defined in terms of three scalars, has a definite physical meaning and hence T cannot be termed as a mere "label" let alone "a bad coordinate". In particular, at $R = \infty$, T is directly the local proper time. Note that in order that $d\tau$ is real, T would become imaginary for $R < \alpha_0$ (if $\alpha_0 > 0$).

Similarly, an infinitesimal displacement in the radial direction ($d\theta = d\phi = dT = 0$), i.e., infinitesimal proper radial length is

$$ds = dl = \frac{dR}{\sqrt{1 - \alpha_0/R}} \tag{33}$$

Thus R is both a scalar and a spacelike coordinate. Again it is trivial to see that in order that dl is real, R would become imaginary for $R < \alpha_0$ (if $\alpha_0 > 0$).

4 Event Horizon

The vector fields normal to a hypersurface $S =$ constant are

$$L = f(x)(g^{ik}\partial_k S)\frac{\partial}{\partial x^i} \tag{34}$$

where f is an arbitrary non-zero function. If L^2 happens to be zero for a particular hypersurface then the corresponding hypersurface is called a "Null Hypersurface"[16]. We shall show afterwards that the Event Horizon marked by $R = \alpha_0$ is such a null surface. For a null surface, the direction of the outward normal coincides with the direction of the tangent and even when photons (or anything else) try to escape the surface, they actually move along the surface and cannot escape the same. Further any light ray emanating in a region $R < \alpha_0$ actually falls inwards, i.e., all surfaces at $R < \alpha_0$ are "trapped surfaces".

Hence no signal or no information can escape out from the $R \leq \alpha_0$ region and therefore the surface $R = \alpha_0$ is called an Event Horizon (EH). This phenomenon is the GR realization of John Michell's primitive idea of a "Dark Star" and much later, in 1967, J.A. Wheeler termed such objects endowed with an EH as "Black Holes". And the name stuck in the psyche of everybody.

Despite this extremely significant *physical property* of an EH which is a surface of *invariantly defined* infinite gravitational redshift (z) and from which even light cannot escape, most of the General Relativists consider the EH as regular region of spacetime without any true physical singularity! On the other hand, the apparent singularity at the EH is considered to be a mere "Coordinate Singularity". But such weird physical properties of the EH can be directly traced to the singular properties of the Hilbert metric coefficients at $R = \alpha_0$. Ironically, in the same breath, most of the GR specialists would often also write "the pull of gravitation is so strong at the EH that nothing, not even light, can escape from within it..". But *if EH is a merely coordinate singularity without any physical significance why does (even) light get trapped by it?* Is not this one of the greatest physical phenomena? Why is the EH a null surface? It seems that neither many are interested in posing this question and nor any body would care to answer this question. On the other hand, almost everybody loves to happily live with the contradictions and mysteries of the concept of a (finite mass) BH.

Even when one would window dress the metric coefficients to make them appear "non-singular" (by resorting to singular transformations) none of the weird physical properties of the EH vanishes and the EH continues to be located by the $R = \alpha_0$ fixed marker.

Yet there are some apparent reasons for which the metric singularity at $R = \alpha_0$ is considered to be a "coordinate singularity" and to appreciate which we should first review some of the basic notions.

4.1 Determinant of the Metric

Consider a Minkowskian spacetime

$$ds^2 = dT^2 - dR^2 - R^2(d\theta^2 + \sin^2\theta d\phi^2) \quad (35)$$

which is necessarily free of any curvature singularity in case it is a global description. Nonetheless

$$g_{\phi\phi} = -R^2 \sin^2\theta \quad (36)$$

has a singularity at $\theta = 0$ (for any R) and both $g_{\phi\phi}$ and $g_{\theta\theta} = -R^2$ have a singularity at $R = 0$. But note that since the choice of the reference line for measuring θ is completely arbitrary, any **non-singular** transformation

$$\theta \rightarrow \theta + \theta_0 \quad (37)$$

would push the singularity to other value of $\theta \neq 0$ and hence this is a genuine coordinate singularity.

Recall that a Minkowskian metric can be obtained when

(i) there is no source of gravitation or when there is no mass-energy at all, i.e., there is no object or no massenpunkt and all curvatures including K are zero (i.e., Riemann =0). In this case, the choice of $R = 0$ is arbitrary, both physically and mathematically, and the corresponding singularity is removable by trivial non-singular transformations. Hence the singularity at $R = 0$, in this case, is a genuine coordinate singularity because it can be shifted to anywhere by making a new choice of the origin of R without compromising on any geometrical advantage.

However,

(ii) the metric (35) could also represent the spacetime in the infinitesimal neighborhood of a source of mass-energy (at $R = 0$) in a locally free falling frame. In this case, neither the choice of $R = 0$ is geometrically arbitrary nor are the curvature components identically zero. In such a case, the singularity at $R = 0$ might be a genuine physical singularity since the singularity persists at the location of the massenpunkt and actually cannot be physically removed by any coordinate transformation.

The determinant of the Minkowskian metric, in spherically symmetric coordinates, is

$$g_M = -R^4 \sin^2\theta \tag{38}$$

And unless one is treading onto $R = 0$ or $\theta = 0$ (where there is no physical singularity), one must have $g < 0$. At $R = 0$, $g_M = 0$ and the occurrence of this null value must be a purely coordinate singularity effect because choice of the origins of both R and θ are completely arbitrary in the absence of any mass-energy, any particle, or any fluid. Thus, in STR, vanishing of g_M is a genuine coordinate singularity effect. Accordingly, despite the vanishing of g_M, there is no compulsion to switch to exotic coordinates in STR, and, in fact, in STR, one can set $g_M = 0$ without any loss of generality, and vanishing of g_M may be seen as absence of any source of gravitation in the problem.

But as soon as one would introduce a single "Point Mass", or a spherical body, mass-energy would be introduced into the problem, and then, in a strict sense, one needs to work with GR instead of STR. And as soon as a spherical distribution of mass-energy is there, there is a natural choice for $R = 0$, if the metric is to reflect the spacetime symmetry in a transparent way.

By Principle of Equivalence (POE), any spacetime may be considered locally flat and the general metric determinant g is related to the corresponding Minkowskian g_M as

$$g = J^2 g_M = -J^2 R^4 \sin^2\theta \tag{39}$$

Hence GR requires that the determinant g of the metric coefficients must be negative or zero; $g \leq 0$, in any coordinate system in order that its sign matches with the corresponding sign of g_M in a free falling Minkowskian spacetime. Here J is the Jacobian corresponding to the transformation connecting the free falling Minkowskian spacetime and the arbitrary coordinate spacetime. Eq.(39) shows that *even when there would be a genuine physical singularity, the singularity in g may remain camouflaged because of covert use of singular* $J = \infty$. Thus finiteness of g need not imply absence a of physical singularity. But vanishing of g in any coordinate system may signify presence of a physical singularity.

By using Eq.(30), it is easily seen that the value of g for the Hilbert metric too is

$$g_H = -R^4 \sin^2\theta \tag{40}$$

Thus not only is g_H negative but also exactly the same as g_M. Remember, however, that, for and extended body of radius R_0, $g_H \neq g_M$ for $R < R_0$, i.e., in the presence of mass-energy; though, at the external vacuum, $R > R_0$, $g_H = g_M$. At the EH, one would have

$$g_H^{EH} = -\alpha_0^4 \sin^2\theta = -16M_0^4 \sin^2\theta \tag{41}$$

and which indeed appears to be negative and non-zero. But note that actually g_H^{EH} would be negative only if one would have $\alpha_0 = 2M_0 > 0$, i.e., one would accept the existence of finite mass point singularities (particles) in GR at the very outset!

(ii) The Kretschmann scalar associated with the Hilbert metric can be seen to be

$$K = R_{iklm}R^{iklm} = \frac{12\alpha_0^2}{R^6} \tag{42}$$

where R^{iklm} is the Riemann tensor. At the EH,

$$K^{EH} = \frac{12}{\alpha_0^4} = \frac{3}{4M_0^4} \tag{43}$$

It appears that K^{EH} remains finite at the EH and it is argued that the EH is not a physical singularity because all true physical singularities are expected to be curvature singularities as well. Note however that K^{EH} would be finite only if one would assume $M_0 > 0$, i.e., accept the existence of finite mass point singularities.

What is conveniently forgotten in this whole exercise is that even if we pardon the assumption of existence of "point singularities", if a region is truly non-singular, then not only the geometric invariants but *none of the physical scalars should blow up there*. But is it really so? Of course it is not so:

This important aspect was first brought out by Abrams[12], Antoci and Liebscher[13] and by Mitra[3]: The norm of the 4-acceleration acting on a test particle

$$a = \sqrt{-a^i a_i} \tag{44}$$

is an Invariant local intrinsic property of spacetime just like the Kretschmann scalar or any other polynomial invariants built using the Riemann tensor. And for a Massenpunkt irrespective of whether one uses the Hilbert metric or Schwarzschild or any related metric, one has

$$a = \frac{\alpha_0}{2R^2\sqrt{1-\alpha_0/R}} = \frac{M_0}{R^2\sqrt{1-2M_0/R}} \tag{45}$$

Obviously this *Physical Invariant representing a local intrinsic property of the space-time* **does blow up** at the EH, $R = \alpha_0 = 2M_0$. Hence the *EH must be a region of true physical singularity*. All those who still claim otherwise must be feigning to be ignorant of this new development in the problem.

It is because of this solid physical reason that Abrams, Antoci and several other authors think that the EH cannot be of finite extent and this must be the site of the Massenpunkt itself. As long as one still sticks to the idea of a finite mass point singularity, this can apparently be accomplished by first (i) rejecting the Hilbert gauge where the Massenpunkt sits at $R = 0$, and on the other hand by accepting original Schwarzschild gauge where (ii) the massenpunkt sits at $r = 0$, i.e., at $R = \alpha_0$.

But we have already discussed the oddities and inconsistencies of this latter solution:

(i) Invariant Circumference of a circle just skirting the Massenpunkt would be $2\pi\alpha_0$ rather than zero which would mean as if the "point mass" has an externally perceived radius of α_0".

(ii) K would still blow up at $r = -\alpha_0$ ($R = 0$) a region where the Massenpunkt is not situated at all

and

(iii) In the presence of a spherical symmetry, the very notion of a region $r = -\alpha_0 < 0$ is an oddity both mathematically and physically; the pulse of light emitted by the massenpunkt would appear to be non-spherical because its intensity would fall off as $(r^3 + \alpha_0^3)^{-2/3}$ rather than as r^{-2}! Thus we reject the original Schwarzschild's viewpoint in this context.

Now going back to Eq.(45), it is seen that if there would be a spacetime beneath $R < \alpha_0$, the Intrinsic Physical Scalar a would become an imaginary quantity after having blown up at $R = \alpha_0$. This situation is just similar to the case where the Lorentz factor of a test particle would be infinite at $v = c$ and then would become imaginary if one would incorrectly assume the existence of superluminal speed $v > c$.

This demands that, there is no spacetime beneath $R = \alpha_0$ and which is possible only if $\alpha_0 \equiv 0$, i.e., if it would be realized that at least in the context of "isolated bodies" GR does not allow the existence of finite mass point singularities. Obviously when $\alpha_0 = 0$, (i) the metric determinant $g_H^{KH} = 0$ (singular and not non-negative) and (ii) $K^{EH} = \infty$ at the EH in conformity with $a = \infty$ and $z = \infty$ at the same region.

Thus although, the Hilbert solution is a correct mathematical solution and best adapted to the spherical symmetry, it apparently leads to the existence of (non-rotating) BHs in GR. But, in order that the associated geometrical and physical scalars diverge *only at a true physical singularity and not at a mere coordinate singularity*, it is necessary that the **undetermined integration constant** $\alpha_0 = 0$. By Birchoff's theorem, Hilbert solution would represent the external spacetime of spherically symmetric extended bodies with radius $R_0 > 0$ as well. The corresponding integration constant $\alpha = 2M$, however, need not be zero, because in such a case, the vacuum Hilbert solutions ceases to be valid for $R < \alpha = 2M$. Thus $\alpha_0 = 0$ does not mean that $\alpha = 2M = 0$ too. In other words, all it means is that such extended bodies must have a circumference radius $R_0 > \alpha = 2M$ and only

$$M_0 = \lim_{R_0 \to 0} M = 0 \tag{46}$$

On the other hand for extended non-radiating static bodies, the Hilbert metric is

$$ds^2 = \left(1 - \frac{\alpha}{R}\right) dT^2 - \left(1 - \frac{\alpha}{R}\right)^{-1} dR^2 - R^2(d\theta^2 + \sin^2\theta d\phi^2); \qquad R \geq R_0 > \alpha \tag{47}$$

4.2 Radial Geodesic Around a Massenpunkt

Here we set, $d\theta = d\phi = 0$ in Eq.(30) so that

$$ds^2 = (1 - \alpha_0/R)\, dT^2 - \left(\frac{dR^2}{1 - \alpha_0/R}\right) \tag{48}$$

Since $\frac{\partial}{\partial T}$ is a Killing vector (metric components do not involve T), we must have energy per unit rest mass[16]

$$E_* = g_{00}\frac{dT}{ds} = (1 - \alpha_0/R)\frac{dT}{ds} \tag{49}$$

as a constant of motion so that

$$\frac{dT}{ds} = \frac{E_*}{1 - \alpha_0/R} \tag{50}$$

Dividing both sides of Eq.(48) by ds^2, we find,

$$1 = \left[(1 - \alpha_0/R) - (1 - \alpha_0/R)^{-1}(dR/dT)^2\right](1 - \alpha_0/R)^{-2}{E_*}^2 \tag{51}$$

or,

$$\frac{dR}{dT} = \frac{1 - \alpha_0}{E_*}[E_*^2 - (1 - \alpha_0/R)]^{1/2} \tag{52}$$

Also,

$$\frac{dR}{ds} = \frac{dR}{dT}\frac{dT}{ds} = \frac{dR}{dT}\frac{E_*}{1 - \alpha_0/R} \tag{53}$$

If the test particle is assumed to be at rest ($dR/dT = 0$) at $R = R_i$ at $T = 0$, it follows from Eq.(52) that

$$R_i = \frac{\alpha_0}{1 - E_*^2} \tag{54}$$

Now introducing the "cycloid parameter" η defined through[17]

$$R = \frac{R_i}{2}(1 + \cos\eta) \tag{55}$$

Eq.(53) can be integrated to find the proper time of infall as ($ds = d\tau$)

$$\tau = \frac{R_i}{2}\left(\frac{R_i}{\alpha_0}\right)(\eta + \sin\eta) \tag{56}$$

Note that the EH corresponds to

$$\cos\eta_H = \frac{2\alpha_0}{R_i} - 1 \tag{57}$$

and

$$\sin\eta_H = \frac{4\alpha_0}{R_i}(1 - \alpha_0/R_i) \tag{58}$$

Under the assumption of $\alpha_0 > 0$, it would appear that the proper time to reach the EH, τ_H, is finite and so is τ_c, the proper time to collapse to $R = 0$. The latter is interpreted as an evidence for sudden termination of the geodesic (since there cannot be any motion beyond $R = 0$) and accordingly an evidence for the existence of a physical singularity at $R = 0$. However the (apparent) finiteness of τ_H and continuation of the geodesic beneath the EH is considered as an evidence that the EH is no physical singularity, i.e., as if, locally, an observer may indeed go inside the EH in a finite proper time.

Note that the above results are obtained by using Eq.(48) and accepting the validity of the coordinates R and T both outside and inside the EH. Note also that nobody here raises the question that R and T cease to become "valid coordinates" or R becomes "timelike" and T becomes spacelike for $R < \alpha_0$. Yet the eventual result of the acceptance of the universal validity of Eqs.(48-58) is consolidation of the idea of an EH beneath which either R and T are supposed to be invalid coordinates or exchange their respective roles!

The Eqs.(52-56) can be integrated to find the coordinate time T of infall[17]:

$$T = \alpha_0 \ln \frac{(R_i/\alpha_0 - 1)^{1/2} + \tan(\eta/2)}{(R_i/\alpha_0 - 1)^{1/2} - \tan(\eta/2)} + \left(\frac{R_i}{\alpha_0}\right)^{1/2} \left[\eta + \left(\frac{R_i}{2\alpha_0}\right)(\eta + \sin\eta)\right] \tag{59}$$

It is seen that unlike τ_H, $T_H = \infty$, i.e., the faraway observer carrying out the experiment never sees the test particle to reach the EH let alone cross the same! Note that T is just no label which can be discarded at inconvenient moments (after using it faithfully for deriving the convenient result that $\tau_c = finite$) because T is the *proper time* of an observer who is carrying out the astronomical observation. In the case of finding τ_c, we had no qualms about using R, T coordinate system inside the EH and if the same viewpoint is retained, we find that for $R < \alpha_0$, T would be an *imaginary* quantity! Earlier too, we found that T must become an imaginary quantity in this regime in order that proper time interval is real. To escape such inconvenient situations some authors quietly put (without any justification) a modular sign around the argument of the logarithmic term in Eq.(59). In such a case T would appear to be a real finite number for $R < \alpha_0$ and in particular as the particle would reach $R = 0$, one would have

$$T = T_c = \pi(R_i - \alpha_0)(1 + R_i/2\alpha_0) \tag{60}$$

But consider the viewpoint of the distant observer as he tracks the infall of the test particle inside the supposed EH:

He finds $T_H = \infty$ as $R = \alpha_0$, and even assuming that the particle crosses the EH, he ought to ascribe $T = \infty$ in such a case too because the observer clock would not get reversed by the particles crossing the EH. Then how can T be finite after already being infinite?

5 Upper Mass Limit of Compact Objects?

One of the most pioneering researches in astrophysics was the realization by Fowler (1926) that upon exhaustion of fuel, there could be compact stellar configurations supported not necessarily by any heat energy but by cold quantum degenerate electron pressure[18]. Such a compact configuration is called a (cold) White Dwarf (WD). By special theory of relativity (STR), the energy of a test particle having rest mass m and momentum p is

$$E = mc^2 \left[\left(1 + p^2/m^2c^2\right)^{-1/2} - 1 \right] \tag{61}$$

irrespective of p being of thermal (collisional macroscopic random motion) or degeneracy (localized non-random non-collisional jittery motion) origin. In the limit of small p, we have

$$E \to p^2/2m \tag{62}$$

Since, in this limit, E increases rapidly with p, the resistive pressure gradient too increases sufficiently rapidly as the star contracts. When p is indeed due to intrinsic microscopic jittery motion around the instantaneous location of the particle itself, it initially appeared that there could be cold WDs of arbitrary high mass and all massive stars (or any massive object) would terminate their evolution into a state of appropriately massive WD.

Anderson (not Chandrasekhar) first considered relativistic effect into Eq.(61); he introduced the concept of "Relativistic Degeneracy"[19] where

$$E \to cp \to \infty \; as \; p/mc \to \infty \tag{63}$$

and Stoner (1929, 1930) [20,21] showed that because of the likely occurrence of extreme relativistic degeneracy with increasing stellar mass M and increasing central density ρ_c, there should be an upper mass limit of (cold) WDs. The basic reason for the occurrence of this limit is that, for extremely large $p \gg mc$, increase in p does not raise E and the pressure gradient sufficiently fast. Chandrasekhar took up these ideas and unified them on a much stronger footing by considering hydrostatic equilibrium of self-gravitating WDs[22,23]. In the limit of Eq.(63), the WD would tend to be a polytrope with $n = 3$. A Newtonian polytrope has[6]

$$M \propto \rho_c^{(3-n)/2n} \tag{64}$$

where ρ_c is the central density, so that as $n \to 3$, M tends to attain a fixed value of

$$M_{ch} \sim \left(\frac{hc}{G}\right)^{3/2} (m_N \mu)^{-2} \sim 5.6 M_\odot \mu^{-2} \tag{65}$$

where m_N is the nucleon mass and μ is the number of pressure giving degenerate particle per nucleon.

For a He WD with $\mu = 2$, one obtains the conventional $M_{ch} \sim 1.4 M_\odot$. Since this theory of WD is entirely Newtonian, it is expected to be valid only at relatively lower values of ρ_c, i.e., for WDs only and not necessarily for configurations with much higher central densities.

The radius of a Newtonian polytrope varies as[6]

$$R \propto \rho_c^{(1-n)/2n} \tag{66}$$

Hence as $\rho_c \to \infty$, $R \to 0$. Thus in reality, Chandrasekhar's critical polytropes are indeed singular configurations with $p \to \infty$, $\rho_c \to \infty$, $R \to 0$, when they are actually meant to be finite objects with density $> \rho \sim 10^6$ g/cm^3 but $\ll \rho_n$, the nuclear density. Further note that, unless n is exactly 3.0, in this limit one would have had $M \to \infty$ if $n < 3.0$. And since approach to such singular situations must be treated by GR rather than by any Newtonian gravity, it would be worthwhile to briefly consider the *General Relativistic Polytropes* of Tooper[24].

Eq.(2.24) of Tooper's paper shows that as $\rho_c \to \infty$, the scale size (radius) of the GTR polytropes $A^{-1} \to 0$ just like their Newtonian counterpart. On the other hand, Eqs. (2.15) and (4.7) of the same paper shows that the gravitational and (not baryonic) mass of the GTR polytropes $M \propto K^{n/2} \propto \rho_c^{-1/2} \to 0$ as $\rho_c \to \infty$ (the K appearing here is not to be confused with Kretschmann scalar). Therefore unlike the Newtonian case, in the case of extreme relativistic degeneracy ($\rho_c \to \infty$), GTR polytropes acquire *zero gravitational mass*. Why does it happen so? This happens because (negative) self-gravitational energy contributes to the gravitational mass in the GTR case while in the Newtonian case gravitational mass is just the bare baryonic mass.

This is not to tell that there cannot be an upper (gravitational) mass limit of *static* and **cold** configurations in GTR. On the other hand, there can indeed be such an upper mass limit but that limit corresponds to some appropriate finite value of ρ_c and not to $\rho_c = \infty$.

To appreciate this subtle point further we shall recall the important work of Oppenheimer and Volkoff (OV) where they found the Upper Mass of a *cold and static* configuration of degenerate neutrons described by an ideal Fermi-Dirac EOS in an *exact* General Relativistic manner[25]. Though technically, this configuration corresponds to $\mu = 1$, the Oppenheimer-Volkoff mass limit so obtained is

$$M_{ov} \sim 0.8 M_\odot \tag{67}$$

rather than $5.6 M_\odot$ suggested by Chandrasekhar's Newtonian work. This huge numerical discrepancy apart, what is of much greater significance here is the fact that *M_{ov} corresponds to a finite value of ρ_c* and not to $\rho_c = \infty$. Thus one must distinguish between two different notions of limiting (gravitational) masses:

5.1 Upper Mass Limit

This is the upper (gravitational) mass limit of a *cold* and static configuration and corresponds to a finite value of the central density (M_{ov}).

5.2 Ultimate Mass Limit

This is the eventual (gravitational) mass limit of a static configuration and corresponds to the singular state of infinite central density. Tooper's work[24] shows that the universal value of this limit is $M_* = 0$ even though the incorrect Newtonian treatment of this problem fudges this limit with the upper mass limit.

Our previous result that as far as Massenpunkt or Point Masses are considered GTR allows only a unique value of $M = 0$ is in perfect agreement with the result that as $\rho_0 \to \infty$, $R \to 0$ and $M \to M_* = M_0 = 0$. However note that mathematical existence of such a limit does not necessarily mean that such a limit is ever attained (in finite proper time of any observer).

At present M_{ov} implies the upper mass limit of any *cold* baryonic configuration described by the most realistic EOS. It is important to note that even if occasionally some authors may consider "finite temperature" effects to improve upon the value of M_{ov}, it still essentially remains the Upper Mass Limit of a *cold* or *nearly cold* baryonic object. Depending on the kind of assumptions and EOSs used the broad range of M_{ov} is[26]

$$M_{ov} \approx (3-6) \times \left(\frac{4.6 \times 10^{14}\ g/cm^3}{\rho_c} \right) M_\odot \tag{68}$$

This expression too shows that if there would be a configuration with $\rho_c \to \infty$, one would approach $M_{ov} \to 0$ in conformity with the result that GTR allows only a unique value of the gravitational mass of a Massenpunkt, namely, $M_* = M_0 = 0$.

6 Collapse of a Uniform Dust

The existence of a Chandrasekhar Limit (more appropriately a Stoner - Chandrasekhar Limit) jolted astrophysicist's complacency about the fate of massive stars upon exhaustion of nuclear fuel. Very massive stars cannot find peace as appropriately massive cold WDs because of the existence of a M_{ch}. While this actually meant that GC of massive stars would continue to an unknown stage beyond modestly massive cold WDs, many authors convey the impression that this "unknown stage" must be a BH. The first person to quietly conceive of BHs in the GC was none other than Eddington. One must bear in mind the fact that this was the time when the existence of a Neutron Star was hardly known (Landau worked out the idea of a Neutron Star in 1932)[27] or Quark Star was unthinkable and absolutely unthinkable was the possibility of truly eternal non-terminating GC. Anyway, at the same time Eddington's great physical intuition goaded him to reject the occurrence of BHs. And this was the reason that Eddington had difficulty in accepting Chandrasekhar's result.

The existence of an upper mass limit of *cold and static* objects in GR too consolidated the idea that sufficiently massive stars must undergo collapse beyond the stages of a WD or a Neutron Star. Oppenheimer and Snyder(OS) wanted to study the evolution of a sufficiently massive self-gravitating configuration to see if a "Dark Star" as suggested by Hilbert (mistaken as due to Schwarzschild) solution for a Point Mass[28].

However as discussed in Section (3), because of the ever evolving nature of the EOS and radiation transport properties, even a Newtonian gravitational collapse cannot be studied exactly by either analytical or numerical methods. Thus a direct *exact* study of genuine continued GC either analytically or numerically is simply an impossibility. As far as analytical study is considered OS realized that they needed to make the drastic assumption of setting the pressure of the fluid $p = 0$. A "fluid" is termed so and is different from a collection of non-interacting incoherent corpuscles because of the macroscopically correlated motions of its different portions and pressure is a manifestation of such macroscopically correlated motion. If pressure would be set to zero, the formalisms developed for correlated fluids would cease to be applicable at all. Yet if one would enforce the fluid formalism in such a case for pure mathematical expediency, then for the sake of thermodynamics, one must set the density $\rho = 0$ too. Because for all physical EOS, the $p \to 0$ state can be attained only by letting $\rho \to 0$. This would be true both in Newtonian physics and GR.

The fact that one must have $\rho = 0$ in this problem of collapse of a dust ball assumed to be in a state of "rest" at $t = 0$ can be seen even without applying any thermodynamics:

In the Newtonian case, if the dust ball is really at "rest" at $t = 0$, then the equation for hydrostatic balance

$$\frac{dp}{dR} = -\frac{GM(r)}{R^2}\rho(0) \tag{69}$$

where $r = R$ is the comoving radial coordinate and not to be confused with the variable r in Section 3. In a static case $r = R$. This foregoing equation directly shows that if $p = 0$, one would have $dp/dR = 0$ and hence $\rho(0) = 0$. Once it is so, it would follow that $\rho = 0$ at any epoch too though we do not require the value of ρ at arbitrary epoch[2,3,29].

Similarly, if one uses OV equation for hydrostatic balance at $t = 0$, one would have

$$\frac{dp}{dR} = -G\frac{M(r) + 4\pi p R^3}{R^2(1 - 2M(r)/R)}[p + \rho(0)] \tag{70}$$

which again leads to $\rho(0) = 0$ if $p = 0$, $dp/dR = 0$.

This result can again be obtained from yet another physical consideration. The sound speed of the fluid is $c_s = \sqrt{dp/d\rho}$. When $dp = 0$, c_s would be finite iff $d\rho = 0$, i.e., iff

$$\rho = \rho_0 = uniform \tag{71}$$

First, this result shows that the idea of a "non-uniform" dust is unacceptable if the dust is to be considered as a fluid. However if the dust is considered as a collection of incoherent particles, and not as a fluid, there could be non-uniform dust (actually, as we will find, the mass of a neutral particle in GR is zero, and hence the mass of any dust shell is zero, and hence dust has uniform density!)

At the boundary of the dust ball, $p_b = 0$, $\rho_b = 0$; but since $p = 0$ in the interior too, p happens to be a continuous function. In this case, in order that c_s too is a smooth continuous and finite function, it is necessary that, $\rho_0 = \rho_b = 0$ too. Thus in any situation, if a dust with $p = 0$ is feigned as a physical fluid, consistency demands that $\rho = 0$ too. In such a case the mass of the spherical fluid of finite extent would be $M = 0$ both in Newtonian and GR case. However, neither OS nor any other GR worker seemed to realize this.

In any case, even by setting $p = 0$, the GR equations for GC cannot be solved uniquely unless one would explicitly incorporate the fact that

$$\rho = \rho_c = uniform \tag{72}$$

With these twin simplifications, it is believed that, OS solved the GC problem analytically and showed the formation of a BH. As discussed in Mitra[2], the metric coefficients do not become singular in the desired manner as the collapse is completed unless one self-consistently considers $M = 0$ for the dust ball. Thus from both physical and mathematical considerations the work of OS describes the formation of a zero mass BH in perfect conformity with the result that $\alpha_0 \equiv 0$.

The proper comoving time for formation of this BH is exactly same as its Newtonian counterpart:

$$\tau_c = \frac{\pi}{2}\left(\frac{3}{8G\pi\rho(0)}\right)^{1/2} = \frac{\pi}{2}\left(\frac{R^3}{2GM}\right)^{1/2} \tag{73}$$

Why is this exact matching? As emphasized before, if any "Dust" is to be used as a fluid in a hydrodynamical equation, then its thermodynamically allowed mass energy density is $\rho = 0$. In the limit of $\rho \to 0$, GR would reduce to Newtonian gravity and this is the physical reason that one obtains exactly the same formula for collapse time both in Newtonian and GR case! Further note that since $\rho(0) = 0$, the proper time of formation of this BH is infinite which means that the collapse becomes truly eternal and even the $M = 0$ BH is never formed!

Nonetheless, if one would consider an inhomogeneous dust, then depending on the density gradient, it may appear that one could obtain either a BH or a Naked Singularity. Such results would have pure applied mathematics appeal and no physical appeal. This is so because, it can be seen that since, physically, one must have $\rho = 0$ for a $p = 0$ EOS, there is no question of having a non-uniform dust model for a *physical fluid*: One can of course have an inhomogeneous dust but such a dust (or even any uniform dust) cannot be conceived as a fluid and used in hydrodynamic equations.

What would, on the other hand, be the proper time of collapse for the case of a fluid with real pressure? Since actual collapse equations cannot be solved, this question cannot be naively answered. However a toy model of collapse of an ideal Fermi-Dirac fluid again leads to $\tau_c = \infty$[3].

How can one in general understand that the collapse to a zero mass BH state is an eternal one with $\tau_c = \infty$. To have a BH, one should be able to reach the limit

$$\lim \frac{2M}{R} = 1 \tag{74}$$

If this is not possible, then the collapse can, at best, asymptotically try to achieve this state, i.e., in actual practice, the foregoing limit would never be achieved. The mathematical way of expressing this impossibility would $\tau_c = \infty$ rather than $\tau_c = finite$. On closer inspection, Eq.(39) of the OS paper does contain this message of impossibility of attaining this limit (74). This Eq. of OS paper is of the form:

$$T \sim \ln \frac{y^{1/2} + 1}{y^{1/2} - 1} \tag{75}$$

where T is the proper time perceived by a distant observer and y is a variable directly related to comoving radial coordinate r :

$$y \equiv \frac{1}{2}\left[(r/r_b)^2 - 1\right] + \frac{r_b R}{2M_b r} \tag{76}$$

Here r_b is r at the boundary and M_b is the mass of the entire dust ball. At the boundary of the dust ball, $r = r_b$ and

$$y_b = \frac{R_b}{2M_b} \tag{77}$$

In order that T is real and definable at all, it is necessary that the argument of the logarithmic term is positive and therefore one must have $y > 1$. In terms of y_b, this directly demands that

$$\frac{2M_b}{R_b} < 1 \tag{78}$$

Thus the dust ball actually never plunges inside its EH and hence one must have $\tau_c = \infty$. In general the condition $y > 1$ can be seen to lead to

$$\frac{2M(r)}{R} < 1 \tag{79}$$

which means that *no trapped surface is ever formed* in the OS problem. We shall show latter that for collapse of isolated bodies this holds true in arbitrary case as well.

Although Einstein did not attempt for any analysis of the OS paper he nevertheless did not believe in the correctness of its claimed result. Immediately after the publication of the OS paper, he correctly argued that if there would be any BH and a test particle would move around the same its speed would exceed c. Hence he insisted that there cannot be any BH[30]:

" The basic result of the study is the distinct understanding that there are no "Schwarzschild singularities" in the real world. Though the offered theory considers only such systems in which particles move by circular trajectories, we have to hardly doubt that the investigation of generic case will lead to the same results. The Schwarzschild singularity is absent because the matter cannot be concentrated in an arbitrary way; otherwise particles composing the clump will reach the speed of light."

But from the very beginning research in GC has proceeded by attempting to solve the field equations by adopting toy EOS without caring much about its physical validity and to study the property of geometrical quantities rather than worrying too much about the very physical ideas which gave rise to this mathematical structure. In this approach, physical quantities like velocity and acceleration hardly ever make their presence felt. In fact, the accurate and general notions of "Proper Distance" "Synchronized Proper Time" needed to arrive at rigorous definitions of velocity, speed, acceleration were probably not adequately developed during Einstein's life time. To the knowledge of the present author, as far as GR literature (in English) is considered, such formal rigorous definitions were first brought out in the textbook "Classical Theories of Fields" by Landau & Lifshitz in 1951. Strangely, despite this development of GR literature in 1951, GR workers continue to ignore or often avoid such notions in their work. Instead many of them prefer to often treat the relevant problems as applied mathematics exercises with favorable assumptions and simplifications. At times they could be extremely reluctant in critically analyzing their assumptions in the light of physical, thermodynamical and even mathematical reasoning if the current paradigm would appear to be endangered by such critical/subtle physical/mathematical reasonings.

Though, to the knowledge of the present author, nobody showed that Einstein's derivation that the speed of the circulating particles would be $v = c$ to be incorrect, his objection to the formation/existence of BHs was nevertheless considered to be pedestrian and was politely ignored. This attitude of ignoring physical reasons for various aspects of BH research became institutionalized over the years following Einstein's death and the current attitude is that the topic of occurrence of BH is beyond any debate. Only discussions which could be allowed in this context seems to be about detection of BHs or topics which would first accept that there are finite mass BHs (or at least naked singularities) in the first place. Any work which does not fall in this line must be dubbed as "obviously incorrect" or ignored or rejected or ridiculed whenever possible. And if there cannot be any professional critique of such distasteful works, they must be attacked absolutely unprofessionally through various internet science fora. For instance the unequivocal reasoning by Abrams, Antoci and others that since the acceleration invariant, which is an intrinsic local property of the metric, blows up at $R = \alpha_0 = 2M_0$, the EH cannot be a mere coordinate singularity has been ignored by most of the working general relativists! Similarly Mazur and Mottola[31] could not publish their idea that GC of massive bodies may result in non-singular "gravstars" rather than BHs in any standard refereed journal even though, occasionally, incorrect papers are published even in most prestigious refereed journals. This is not to tell that the paper by Mazur and Mottola is correct; in fact, towards the end of the review, we would briefly point out that the entire scheme of "Dark Energy Stars" is unphysical and incorrect. The proofs by Mitra[2,3] and Leiter and Robertson[32] that trapped surfaces or EHs cannot form in order that the timelike radial geodesic of a test particle remain always timelike (at a non-singular region) has similarly been ignored by most of the working relativists including the proponents of "Dark Energy Stars" even though the concepts of "gravstars" and "Dark Energy Stars" came following our proof of non-existence of BHs.

7 More on Hilbert BH

If one persists with the assumption of existence of finite mass Massenpunkts (Point Singularity) then the existence of a BH with its EH at a finite $R = \alpha_0$ would appear to be a reality. And BH research has so far progressed by assuming the existence of point singularities in the first place. From the very beginning, to accommodate this physical incongruity, BH research has used one mathematical manipulation after another to search for a workable framework. We again recall that GR starts with the notion that, in principle, one can use arbitrary coordinate system to analyze any problem though symmetry adapted coordinates would obviously simply the mathematics[8,9,17,33,34]. Thus in GR, except for likely mathematical simplifications and transparent physical explanations, all coordinate systems are on equal footing. There are of course some trivial "coordinate singularities" like vanishing of $g_{\phi\phi}$ in polar coordinates. But such trivial singularities are present both in Newtonian and relativistic physics and, further, formal existence of such trivial singularities need not debar use of polar coordinates. For instance, one gets the same physical result both in polar and Cartesian coordinates despite the presence of such "coordinate singularities" in polar coordinates. Mere use of polar coordinates would never give rise to physically anomalous measurable results such as occurrence of infinite redshift, infinite Lorenz factor, infinite energy or infinite acceleration. Thus as per GR, no coordinate system can give an anomalous physical picture unless there is an intrinsic anomaly in the problem itself.

This basic premises of equivalence of all coordinate systems was the first casualty in order to uphold the incorrect Newtonian assumption of existence of point singularities (i.e., BHs). Thus the singularities in the metric coefficients were ascribed to the usage of "faulty" or "bad" coordinates R. This was indeed unfortunate because R is not only a *scalar*, most adapted to the symmetry of the problem ($2\pi R$ is the invariant circumference of circles around the center of symmetry) but also $R \propto K^{-1/6}$, where K is the Kretschmann scalar. Also the time $t = T$ appearing in the Hilbert metric is the *proper time* of a distant observer. Then why should R and T be bad coordinates particularly when, in GTR, there are really no "bad" or "good" coordinates?

The vacuum Hilbert metric coefficients g_{TT} and g_{RR} do exchange their respective signatures if there would be a region with $R < \alpha_0$; this itself is a great oddity and is usually explained away by postulating that R become "timelike" and T becomes "spacelike" for $R < \alpha_0$. In another version, one would postulate that R and T coordinates breakdown for $R < \alpha_0$ and one must use appropriate new coordinates to describe the region beneath the EH. In this context, we may recall the comments by Rosen[35] :

"..so that in this region R is timelike and T is spacelike. However, this is an impossible situation, for we have seen that R is defined in terms of the circumference of a circle so that R is spacelike, and we therefore are faced with a contradiction. We must conclude that the portion of the spacetime corresponding to $R < 2M$ is non-physical. This is a situation which a coordinate transformation even one which removes a singularity cannot change."

Actually what happens is that T becomes imaginary rather than usurp the place of R. From Eq.(32) it is seen that in order that $d\tau$ remains real, T has to be imaginary for $R < \alpha_0$. Occurrence of such an imaginary T simply indicates that the region $R < \alpha_0$ cannot exist or be born.

Similarly, from Eq.(33), we see that, in order that the *proper radial distance* interval dl remains a real quantity, it follows that R has to be imaginary for $R < \alpha_0$! Further the denominator of the same Eq. demands that if *R would be imaginary, so must be α_0 unless* $\alpha_0 \equiv 0$ in the first place.

Despite this, there have been systematic attempts at inventing "good coordinates" to describe the spacetime beneath the EH of a supposed finite mass point singularity by discarding R and T which have enormous physical and geometrical significance. One must wonder if the choice of coordinates is indeed arbitrary why should R and T fail to describe the region $R < \alpha_0$ even if we ignore their geometrical and physical significance?

7.1 Extended Hilbert Manifold

If one would cut out a patch from an originally continuous sheet of paper, the edges of the patch may appear to be discontinuous and singular. And once one would put that patch back in its original location, the singularities at the edge would vanish. Mathematical relativists and topologists have been trying to explain away the singularity at the EH by the above mentioned line of thinking without telling why should a certain region of the spacetime would appear to behave in this anomalous manner when the choice of coordinates is completely arbitrary in GTR. Several new coordinate systems were invented for over 40 years (approximately from 1921-60) in search of this supposed "cut out" part which would join smoothly to the Hilbert metric at $R > \alpha_0$. In other words, one looks for supposed maximal extension of Hilbert manifold and it is believed that this odyssey got completed in 1960 with the invention of the Kruskal coordinates.

One of the very first attempts for such good coordinates, ironically, is due to none other than Eddington who would later refuse to accept the idea of a BH and accordingly reject Stoners's and Chandrasekhar's discovery of an upper mass limit of cold WDs. Since then Finkelstein, Wheeler, Synge, Kruskal and many other authors have contributed to this attempt. As a purely mathematical exercise and without any intention of establishing the concept of a BH, Eddington noted that one could pass beyond the edge of the Hilbert chart by introducing a R, u coordinate system (θ, ϕ unchanged) where u is the retarded time

$$u = T - \int_0^R \frac{dR}{1 - \alpha_0/R} \tag{80}$$

However for a proper appreciation of this topic, we would deviate here from a chronological account. Since light rays follow null geodesics, from Eq.(48), we see that for radially moving photons,

$$dT = \pm \frac{dR}{1 - \alpha_0/R} \tag{81}$$

The time taken by a photon to travel from the origin to $R = R$ is thus

$$T = \pm \left[R + \alpha_0 \ln \left(\frac{R - \alpha_0}{\alpha_0} \right) \right] = \pm R_* \tag{82}$$

Here R_* is known as the Regge-Wheeler coordinate and plays the key role in the development of "Null Coordinates" i.e., coordinates related to motion of photons,

$$u = T - R_* \tag{83}$$

and

$$v = T + R_* \tag{84}$$

Here u and v are outgoing and ingoing null coordinates respectively. So Eddington's time coordinate (Eq.80) was just this outgoing null coordinate in terms of which

$$ds^2 = (1 - \alpha_0/R)du^2 + 2dRdu - R^2 d\Omega^2 \tag{85}$$

where $d\Omega^2 = d\theta^2 + \sin^2\theta d\phi^2$. The metric now appears to be regular in the sense that none of its (covariant) coefficients diverge at $R = \alpha_0$.

However note that actually $g_{uu} = 0$ at $R = \alpha_0$ and then it changes its sign as it happens for g_{TT}. The metric determinant appears to be finite and negative (for $\alpha_0 > 0$) at $R = \alpha_0$; but we have already mentioned that *a singularity in J may camouflage the singularity in g*. Because of the presence of the $dRdu$ term, the metric now loses its time inversion symmetry even though the progenitor Hilbert metric has this symmetry. Instead of looking for time inversion symmetry in metric(85), here one introduces a mirror metric

$$ds^2 = (1 - \alpha_0/R)dv^2 - 2dRdv - R^2 d\Omega^2 \tag{86}$$

which in itself, again, lacks the time inversion symmetry. Metric (86) is ascribed to outgoing geodesics. It is believed that u and v are well behaved good coordinates varying smoothly over $-\infty$ to $+\infty$. Is it really so?

Let us simply consider the Regge-Wheeler coordinate which is the basis of null coordinates:

$$R_* = R + \alpha_0 \ln\left(\frac{R - \alpha_0}{\alpha_0}\right) \tag{87}$$

Well, $R_* = +\infty$ at $R = \infty$ and $R_* = -\infty$ at $R = \alpha_0$ (if $\alpha_0 > 0$). But, clearly R_* is definable only over the range $\alpha_0 \leq R \leq \infty$ and contrary to the claim made so far, R_*, actually, cannot cover the region $R < \alpha_0$. This can be seen explicitly by noting that R_* **would be imaginary for** $R < \alpha_0$:

$$R_* = Imaginary; \qquad R < \alpha_0; \;\; unless\; \alpha_0 = 0 \tag{88}$$

Similarly, actually both u and v are imaginary for $R < \alpha_0$ just like T and R unless $\alpha_0 = 0$! Again to hide this inconvenient physical and mathematical truth, many authors quietly put a modular sign in the argument of the logarithmic term of R_* without even mentioning so:

$$R_* = R + \alpha_0 \ln\left|\left(\frac{R - \alpha_0}{\alpha_0}\right)\right| \tag{89}$$

For instance, while Misner, Thorne & Wheeler (MTW) did not originally obtain any modular sign in the definition of R_* (see Eq.[25-3] in p. 663), they put a modular sign in Box 31.2, p. 828! Following them, many other authors have practised this academic duplicity. Now supposing for a moment that $\alpha_0 > 0$ and by accepting this academic duplicity of imposing a modular sign for the time being, we would find that, at a region below the EH, such as at $R = 0.5\alpha_0$, R_* would be positive definite when it was already $-\infty$ at $R = \alpha_0$. This is definitely no example of smooth behavior of R_* and associated null coordinates.

Further see that in this case, metric determinant,

$$g = -R^4(1 - \alpha_0/R)\sin^2\theta \tag{90}$$

is zero not only at $R = 0$, the true singularity, but also at $R = \alpha_0$. This strongly suggests, $\alpha_0 = 0$.

Note that since the Regge-Wheeler and null coordinates are obtained by integrating the vacuum Hilbert metric all the way from $R = 0$, they are not valid for describing the spacetime of finite bodies having $R_0 > 0$. Even the external vacuum spacetime of such extended real bodies cannot be described by such coordinates because they already presume that the entire region above $R > 0$ is a vacuum.

In any case, by using the null coordinates one can formally see that the EH is a null surface in the following way:

7.1.1 Null Surface

In terms of the R, v coordinates, the vector field normal to the hypersurface $S = R - \alpha_0 = 0$ is

$$N = f(R)\left[(1 - \alpha_0/R)\frac{\partial S}{\partial R}\frac{\partial}{\partial R} + \frac{\partial S}{\partial R}\frac{\partial}{\partial v} + \frac{\partial S}{\partial v}\frac{\partial}{\partial R}\right] \tag{91}$$

$$= f(R)\left[(1 - \alpha_0/R)\frac{\partial}{\partial R} + \frac{\partial}{\partial v}\right] \tag{92}$$

and

$$N^2 = g^{ik}\partial_i S \partial_k S f^2 = g^{RR} f^2 = (1 - \alpha_0/R) f^2 \tag{93}$$

Hence $N^2 = 0$ at $R = \alpha_0$ and the EH is a null surface. So light cannot escape out of the EH and it is a one-way membrane. This would, however, be in violation of the time inversion symmetry of Einstein equations. And hence the question arises again, can an EH ever form?

7.2 Eddington -Finkelstein Coordinates

Further modifications to Eddington coordinates were carried out by Finkelstein in 1958[37]. Rather than using the Regge Wheeler coordinate R_*, he instead employed a coordinate

$$T_* = T \pm \alpha_0 \ln\left(\frac{R - \alpha_0}{\alpha_0}\right) \tag{94}$$

and retained R as the radial coordinate. Like the null coordinates, Eddington coordinates too can be used only to describe the spacetime associated with a massenpunkt where the entire region $R > 0$ is free of any mass energy. Later this coordinate system came to be known as "Eddington- Finkelstein" coordinates in terms of which we have

$$ds^2 = \left(1 - \frac{\alpha_0}{R}\right) dT_*^2 \pm \frac{2\alpha_0}{R} dT_* dR - \left(1 + \frac{\alpha_0}{R}\right) dR^2 + R^2 d\Omega^2 \tag{95}$$

Like the Eddington metric, Finkelstein metric too suffers from time asymmetry. Finkelstein worried a lot about this time asymmetry and in desperation ascribed this to "non-linear" nature of gravitation forgetting that Einstein equations too are non-linear yet time symmetric. Further, since in the present case, one is handling a "one-body" problem rather than a many-body problem, there is no chance of having a thermodynamical arrow of time. This time asymmetry can actually be removed only by realizing that $\alpha_0 \equiv 0$.

7.3 Lemaitre Metric

If a free falling test particle could really cross the EH, it would be reasonable to use a coordinate system in which the time coordinate is just the comoving time t of the particle. Since for a free falling particle comoving time is the proper time, t is also the comoving proper time[9,33,38]:

$$dt = dT + \frac{\sqrt{\alpha_0/R}}{1 - \alpha_0/R}\, dR \tag{96}$$

and

$$t = T + 2\sqrt{\alpha_0/R} + \alpha_0 \ln\left(\frac{\sqrt{R} - \sqrt{\alpha_0}}{\sqrt{R} + \sqrt{\alpha_0}}\right) \tag{97}$$

where $R_i = \infty$ is assumed. The comoving radial coordinate is defined by

$$dr = dt + \sqrt{R/\alpha_0}\, dR = dT + \frac{\sqrt{R/\alpha_0}}{1 - \alpha_0/R}\, dR \tag{98}$$

so that

$$R = \left[\frac{3}{2}(r - t)\sqrt{\alpha_0}\right]^{2/3} \tag{99}$$

In terms of these comoving coordinates, Lemaitre metric, for a massenpunkt, becomes

$$ds^2 = dt^2 - (\alpha_0/R)dr^2 - R^2 d\Omega^2 \tag{100}$$

On the other hand for a finite body with radius $R_0 > \alpha = 2M$, the Lemaitre metric is

$$ds^2 = dt^2 - (\alpha/R)dr^2 - R^2 d\Omega^2; \qquad R > R_0 = \alpha \tag{101}$$

Lemaitre metric looks like a non-stationary one although actually the field of a massenpunkt is a static one. Thus *explicit appearance of time coordinate in the metric coefficients need not imply existence of non-stationary spacetime*; if the time dependence is non-removable by any coordinate transformation, only then, the spacetime may be considered to be genuinely non-stationary.

As to the metric for the finite body, there is obviously no singularity as the metric is not applicable for $R < \alpha$. And at no value of R, does this free falling metric become completely flat except, of course, in the infinitesimal neighborhood of the particle because $\alpha/R \neq 1$. But for the BH case(Eq.[99]), $g_{rr} = -1$ at the EH so that

$$ds^2 = dt^2 - dr^2 - R^2 d\Omega^2; \qquad R = \alpha_0 \tag{102}$$

This shows that **as if the entire EH would be at the origin of the free falling coordinate system** once the particle would reach the EH. And this is possible, only if the *EH itself is the location of the massenpunkt*, i.e., $\alpha_0 = 2M_0 = 0$ just like $R_0 = 0$.

7.4 Kruskal -Szekeres Coordinates

It is mentioned in the literature that even the Eddington- Finkelstein coordinates cannot give an adequate one-piece description of the maximal Hilbert spacetime. On the other hand this feat is supposed to be fulfilled with the invention of a coordinate chart independently by Kruskal and Szekeres[39,40]. They achieved this by means of the following coordinate transformations:

For the exterior region (Sector I):

$$U = f_1(R) \cosh\left(\frac{T}{2\alpha_0}\right); \qquad V = f_1(R) \sinh\left(\frac{T}{2\alpha_0}\right); \; R \geq \alpha_0 \tag{103}$$

where

$$f_1(R) = \pm\left(\frac{R}{\alpha_0} - 1\right)^{1/2} e^{R/2\alpha_0} \tag{104}$$

And for the region interior to the horizon (Sector II), we have

$$U = f_2(R) \sinh\left(\frac{T}{2\alpha_0}\right); \qquad V = f_2(R) \cosh\left(\frac{T}{2\alpha_0}\right); \; R \leq \alpha_0 \tag{105}$$

where

$$f_2(R) = \pm\left(1 - \frac{R}{\alpha_0}\right)^{1/2} e^{R/2\alpha_0} \tag{106}$$

The $\pm$ sign in f_1 and f_2 refer to "other universes" associated with the maximally extended spacetime of a massenpunkt.

Given our adopted signature of spacetime (-2), in terms of U and V, the metric for the entire spacetime is

$$ds^2 = \frac{4\alpha_0^3}{R} e^{-R/\alpha_0}(dV^2 - dU^2) - R^2(d\theta^2 + d\phi^2 \sin^2\theta) \tag{107}$$

Clearly, we have

$$-g_{UU} = g_{VV} = \frac{4\alpha_0^3}{R} e^{-R/\alpha_0} = \frac{4R_g^3}{R} e^{-R/R_g} \tag{108}$$

The metric coefficients *appear* to be regular everywhere except at the intrinsic singularity $R = 0$, as is expected. These supposed "good coordinates" or rather "best coordinates"

U and V however have no direct physical significance as they are neither scalars nor any measurable quantities. The same is true for R_*, T_*, u v and all such exotic coordinates. Further all such good or better or best coordinates are made up of none other than the "bad" Hilbert coordinates R and T which are supposed to break down within the EH! Is this a logical and consistent approach? In fact, we have already seen that actually R_* and all associated coordinates are definable only for the $R \geq \alpha_0$ region. All such coordinates too become imaginary just like R and T if one would imagine a spacetime with $R < \alpha_0$. Further $ds^2 = 0$ at the EH in all coordinates indicating all geodesics must terminate right at the EH. Note that, the angular part of the metric remains unchanged by such transformations and R actually continues to signal its intrinsic spacelike nature.

As to the Kruskal metric, what is conveniently overlooked in this excitement of having apparently found a one-piece maximally extended time-symmetric Hilbert spacetime is that

the fundamental physical requirement of asymptotic flatness of the spacetime around a Massenpunkt gets abandoned! At $R = \infty$, $g_{UU} = g_{VV} = 0$ rather than being unity! Thus, in reality, the Kruskal coordinate system is the worst coordinate system ever conceived and is just unfit to describe any physical scenario. But does this statement not contradict the basic premises that there is really no "bad" coordinate system? It does not, because, when one conceives of a coordinate system, one allows the coordinates to occupy all likely range of values. But this is violated for the Kruskal case: It is the only coordinate system where one divides physically meaningful quantities like R and T by α_0, i.e., by 0! This is the basic reason that Kruskal coordinate gives rise to most absurd result like the speed of a radially moving test particle, $v = 1$ at $R = \infty$[41,42]!!! This result has recently been found by Burghardt too by using Differential Geometry[42].

Naturally, it is this Kruskal metric which has either given rise to or consolidated some of the most unphysical or science fiction concepts like "White Holes", "Worm Holes" and "Time Machines". Thus we would not anymore deal with Kruskal coordinates in this review. It may also be mentioned that the Ref.[42] has shown that in the Painleve coordinates too, one has $v^2 = 1$ and $ds^2 = 0$ at the EH.

8 More on Coordinate Singularity

In Section 4 we have already seen that a genuine coordinate singularity can always be removed at a given region and pushed to another region by suitable **non-singular** spacetime transformation. Moreover, it is really not necessary to remove any coordinate singularity at all, they are totally innocuous. And none of the physical results get affected by the presence of mere coordinate singularities. For instance, the presence of singularities in $g_{\phi\phi} = -R^2 \sin^2\theta$ does not render any measurable quantity, such as, angular velocity, Lorentz factor, acceleration, pressure or density infinite at $\theta = 0$. Similarly, the mere presence of coordinate singularities in $g_{\phi\phi}$ and $g_{\theta\theta}$ at $R = 0$ does not make such other physical parameters diverge there. Such likely divergence at $R = 0$ would nevertheless may occur due to occurrence/formation of point mass singularities at $R = 0$. And if one would place the point mass singularity at $R = R_0'$ then the coordinate singularity at $R = 0$

would be removed.

Thus the original notion of spacetime singularities was associated with occurrence of singularities in one or more components of the metric excluding those trivial coordinate singularities in $g_{\theta\theta}$ and $g_{\phi\phi}$. In the following we shall attempt to show that once such non-trivial metric singularities are present in a given coordinate system they are *non-removable unless one adopts a* **singular** coordinate transformation.

Let us first consider an example of supposed "coordinate singularity" provided by Wald:

$$ds^2 = \frac{1}{t^4}dt^2 - dx^2 \tag{109}$$

where x is defined over $-\infty < x < \infty$. This metric appears to be singular at $t = 0$. But it appears to be removable by a coordinate transformation

$$t \to t' = \frac{1}{t} \tag{110}$$

by which the metric becomes the flat Minkowskian metric

$$ds^2 = (dt')^2 - dx^2 \tag{111}$$

And now there is no metric singularity and hence Wald argues that metric singularities are no indication of genuine physical singularity. But note that the transformation (109) itself is **singular** with $J = \infty$ at $t = 0$ where the metric was initially appearing to be singular. Had Wald used any other transformation which would not be singular at $t = 0$, the *metric singularity would have persisted*.

Another example of oft-mentioned supposed non-trivial coordinate singularity is that of Rindler Metric[16,43]

$$ds^2 = x^2 dt^2 - dx^2 \tag{112}$$

with coordinate range $-\infty < t < \infty$, $0 < x < \infty$. Though there is no spacetime curvature associated with this metric, the metric shows a singular behaviour at $x = 0$. By a series of coordinate transformations whose eventual form is

$$x = (T^2 - X^2)^{1/2}; \qquad t = \tanh^{-1}(T/X) \tag{113}$$

metric (111) assumes the familiar form of

$$ds^2 = dT^2 - dX^2 \tag{114}$$

Note that one obtains the non-singular metric (113) out of the apparently singular metric (111) by means of a coordinate transformation which becomes singular at $x = 0$ ($t = \pm\infty$). Thus singular transformations are double edged swords which can (apparently) remove genuine singularities or artificially inflict singularities when there is no genuine singularity. So it follows that if there is no metric singularity in a given coordinate system (except trivial polar singularities) the metric would always remain non-singular unless one covertly forces singular coordinate transformations onto it. And if a metric singularity (non-singularity) appears in a physical problem (not in concocted applied mathematics examples) in a metric involving coordinates having physical identification then it is expected that there is a genuine physical (non-physical) singularity.

The question now arises is that if one is unaware of the underlying singular coordinate transformation connecting the two set of coordinates, how would one know in the first place whether there is a spacetime singularity or not? Obviously one must try to evaluate the Kretschmann scalar K in such a case. If K is unambiguously finite, then of course, there is no genuine singularity. However if finiteness of K is subject to the finiteness of other quantity (such as α_0), then, one cannot arrive at a definite conclusion.

Moving away from the purely mathematical notion of arbitrary coordinate systems, in physical world, when one handles a physical problem, the coordinates often are related to measurable physical parameters such as "rod length" "proper time" of some clock, some scalar, behavior of light pulse (intensity falling off as R^{-2}), red shift etc. If the t, x coordinates are made of such measurable physical quantities, a metric like (108) may actually arise in a physical problem.

But the fact is that, a metric such as (108) never appears directly in the context of a physical problem. On the other hand, the t', x coordinates have direct physical meaning; in the absence of negligible gravitation, in a 1-D problem dt' is the proper time of some observer and dx is the proper length of some rod. In physical situation, metric (110) does arise and the example (108) has been concocted by going backward from the known physical metric (110) only to illustrate the notion of "coordinate singularity at the EH".

Again note that coordinates x, t appearing in metric (111) may not have any direct geometrical or physical significance while X, T do possess such physical significance. And actually, one arrives at metric (111) by starting from the known physical metric (113) involving physically meaningful coordinates and then by working backward involving a coordinate transformations which is singular at $x = 0$

$$t = \pm\infty; \qquad at\ x = 0 \tag{115}$$

If X and T are interpreted as the spatial and temporal coordinates of 1-D STR motion, then, $x = 0$ or $X = T$ would correspond to attaining a physical speed of $v = c$ and Lorentz Factor $\gamma = \infty$. Thus singularity in the Rindler's metric corresponds to a supposed infinite acceleration of a test particle in the absence of gravity (no spacetime curvature). In other words, a supposed non-singular physical situation ($v < c$) is pushed unrealistically (by camouflaging it in singular coordinate transformation) to force a singularity in the metric (111).

Further, note, occurrence of metric singularities (except polar type trivial ones), irrespective of whether they are removable or not, corresponds to some sort of unphysical situations. For example, if the metric is to be used to describe the timelike motion of a material test particle, we must have $ds^2 > 0$. But since $t' = \infty$ corresponds to an extremum of t', we would have $dt' = 0$ at $t' = \infty$ or at $t = 0$, so that at $t' = \infty$, we have

$$ds^2 = -dx^2 = 0 \tag{116}$$

because $x = 0$ is an extremum of x. With reference to Eq.(108), since ds^2 must be invariant, dt^2/t^4 must be 0 at $t = 0$. In any case, the metric loses it timelike character at the so-called "coordinate singularity" at $t = 0$. This strongly suggests that unless one constructs highly contrived applied mathematics brain teasers, probably metric singularities indeed correspond to physical anomalies even when some (singular) coordinate transformation may window dress the metric to look "singularity free".

8.1 Coordinate Singularity in Hilbert Metric?

In any physical problem involving differential equations, the integration constants are to be fixed by appropriate physical or initial conditions. In Nature, all bodies are non-singular and could be of finite extent. The concept of a "point" is only a purely geometrical one. Thus, in the context of finite spherical bodies, it is legitimate to identify the integration constant α as $2M > 0$ by using Kepler's 3rd law which gives the definition of gravitational mass. But, in the context of a massenpunkt, since, in Nature, there is no uncharged "point mass", Kepler's 3rd law or any other law cannot be naively used in fixing α_0; and even if such a fixing is to be made, one cannot a priori assume $M_0 > 0$. At best one needs to explore the value of M_0 by pursuing a limit. Suppose we let R_0 decrease by keeping density profile intact, and in this case the value of α would decrease. What would be the value of

$$\alpha_0 = \lim_{R_0 \to 0} \alpha? \tag{117}$$

Physically it could very well be zero but all along we presume $\alpha_0 > 0$, i.e., we *assume the existence of neutral point singularities* following our Newtonian hangover and it is this notion which has given rise to BHs.

Had we kept α_0 as an undetermined quantity and demanded that no Lorentz factor, no Acceleration, no physical scalar, no geometrical scalar would blow up at a region which is not the site of the massenpunkt, we would have been able to correctly fix the value of α_0. But this has rarely been attempted and the research has progressed in the following vicious cycle

(a) First assume that $\alpha_0 = 2M_0 > 0$, i.e., assume that the massenpunkt is a point singularity with infinite density as would be the case in Newtonian physics.

(b) Now calculate $g^{EH} = -{\alpha_0}^4 \sin^2\theta = -16{M_0}^4 \sin^2\theta$ at the EH, which appears to be finite and negative. Though this conclusion is an artifact of the assumption $\alpha_0 > 0$, i.e., existence of finite mass BHs, use this *apparent* result to justify the assumption $\alpha_0 > 0$.

(c) Similarly calculate the Kretschmann Scalar $K^{EH} = (3/4M_0^4)$ at the EH; feed in the assumption $M_0 > 0$ to conclude that BHs are real. Then again use this *apparent* result $K^{EH} = finite$ to justify the original assumption of $\alpha_0 = 2M_0 > 0$.

(d) Having completed this vicious circle of logic for the existence of BHs, do stick to it and bypass or obfuscate any physical difficulty arising in this paradigm

1. By arguing for complex topological spaces which need not have any relation to measurable physical quantities

2. Even if any coordinate having direct geometrical or physical significance would behave in an anomalous way or even if they would become imaginary, try to put a modular sign at appropriate place to suppress such an anomaly.

3. If 3-speed of a material particle would become equal to the speed of light at the EH, in violation of the basic premises of STR and GTR and even if *covariantly defined*[33,43,44] and directly measurable $z = \infty$ at the same place, then also do not admit the EH as a physical singularity. On the other hand, try to obfuscate the measurable physical anomalies by assigning them to "coordinate singularities" with an air of profound thinking. Also conveniently forget that in GTR, mathematically, all coordinate systems are on equal footing and hence no coordinate system can yield a $v = c$ or $z = \infty$ result (which is in any case invariantly defined) unless there is a physical singularity:

For instance while MTW calculate the radial velocity of a free falling observer, they first calculate the velocity of the observer with respect to an observer at a fixed $R = R$[17] (p. 821) :

$$v^R = -\sqrt{\frac{2M}{R}} \tag{118}$$

This shows that in order to calculate quantities in the free falling observers rest frame one needs a "boost" whose value can exceed c. However rather than seriously worrying about the origin of this super luminal boost they bypassed the issue by vaguely writing[17] (p. 821):

"The amazing result (a consequence of special algebraic properties of the Schwarzschild geometry, and somewhat analogous to what happens - or, rather does not happen - to the components of the electromagnetic field **E** and **B**, when they are both parallel to a boost) is this:.."

4. Ignore any result which may be uncomfortable, for instance, ignore the result, that physically measurable local *Acceleration Invariant* diverges at the EH or a supposed timelike metric becomes null at $R = \alpha_0$.

Similarly ignore the result that an initially timelike radial geodesic of a test particle would appear to be null at the EH in violation of the basic premises of GTR.

9 Speed in Special Relativity

We briefly review this apparently trivial topic to highlight some of the often forgotten basic results. Consider the motion of a material test particle in the 2-D spacetime where the signature of the full 4-D spacetime is -2:

$$ds^2 = dt^2 - dx^2 = dt^2(1 - v^2) \tag{119}$$

where $v = dx/dt$ is the speed of the particle measured by an observer O at a fixed $x = x$ or by any other observer O' at relative rest w.r.t. O (in contrast, in GTR, one needs to define v

only locally). The fundamental requirement that $ds^2 > 0$ (for a spacetime signature of +2, one would instead have $ds^2 < 0$) for the timelike worldlines of the material particle directly demands that

$$v^2 < 1 \tag{120}$$

Historically Eq.(119) was first postulated guided by Michelson- Morley experiment, and the more abstract and general concept of $ds^2 > 0$ came later. On the other hand, in a much more complex spacetime embedded by gravitation the abstract notion of "time like worldline" with $ds^2 > 0$ becomes more suitable. However this does not mean that the notion of a 3-speed is no more tenable in GTR as many authors would like to project in order to retain the BH paradigm.

Thus when it is mentioned that "a timelike worldline must always remain timelike" it is implied that one would always honor and ensure the validity of $v^2 < 1$ for the material particle. However if one would artificially conceive of imparting infinite acceleration and Lorentz factor to the particle, i.e., misconceive of a physical situation where $v = c$, obviously then one would have

$$ds^2 = 0 \tag{121}$$

Alternatively, an occurrence of $ds^2 = 0$ situation cannot be ignored by insisting that one must have $ds^2 > 0$ without realizing that one might indeed be pushing a physical problem ($v^2 < c^2$) beyond its limit to a singular $v = c$ situation. The occurrence of $v = c$ or $ds^2 = 0$ thus must be related to injection of some sort of physical singularity into the problem. As such the occurrence of $ds^2 = 0$ does not violate the rule that the norm of the four velocity (when defined in terms of s) must always be same $(+1)$ because

$$u_k u^k = g_{ik} u^i u^k = g_{ik} \frac{dx^i}{ds} \frac{dx^k}{ds} = \frac{ds^2}{ds^2} = 1 \tag{122}$$

is an identity and valid irrespective of the value or sign of ds^2.

Moreover since ds^2 *is an invariant*, once $ds^2 \to 0$, the situation cannot be remedied by any coordinate transformation, not even those involving singular Jacobian J. Consequently, it follows from Eq.(118) that once $v \to 1$ in any coordinate system one must have $v' \to 1$ in any other coordinate system too even though when $v < 1$, $v \neq v'$. This physical requirement is ensured in the well known STR velocity addition law:

$$v' = \frac{v + v_0}{1 + v v_0} \tag{123}$$

where v_0 is the relative velocity (taken as an algebraic quantity) between O and O'.

It may be also noted that as long as the acceleration a is finite in one frame, the transformed acceleration $a' \neq a$. But once $a = \infty$ in one frame, then STR acceleration transformation rule shows that all transformed acceleration $a' = \infty$ too.

Now returning to Rindler's metric, from Eq.(112), we see that, $x = 0$ corresponds to $X = T$. Then by differentiating the same equation, it is found that at $x = 0$, we have

$$v = \frac{dX}{dT} = \frac{T}{X} = 1 \tag{124}$$

Therefore, at $x = 0$, the metric (113) yields $ds^2 = 0$ although for $x \neq 0$, one has $ds^2 > 0$. Thus again we see that metric singularity means serious physical anomaly where the metric may lose its time like character. And the metric singularity may correspond to the unphysical situation of attainment of infinite Lorentz factor and acceleration.

Therefore referring to original Rindler metric, again, one would always have $ds^2 > 0$ if one would not unphysically push the situation to $x = 0$ (i.e., $v = 1$). And thus there must not be any attempt to extend the Rindler manifold to $x = 0$ and beyond by physically untenable applied mathematics exercise. On the other hand, the manifold must exclude the STR singularity $x = 0$ if $ds^2 > 0$ is to be honored.

Since STR is supposed to be locally valid in GTR too, these basic results must be true in GTR too provided all the quantities are measured in the same location (i.e. at same R, θ, ϕ).

9.1 Speed using Comoving Coordinates

In the above discussions, it was implicitly assumed that the coordinates x, t referred to fixed coordinates, as if x refers to a milestone on a road and t refers to the time measured by a clock fixed in that milestone. Here v could be interpreted as the speed of a moving car passing by the milestone. But, now suppose, both x and t are comoving coordinates, i.e., both the observer and the clock is fixed in the car rather in the milestone. Since x is fixed, $dx = 0$ as reckoned by the comoving observer. Then does it mean that, the metric would, in general, be reduced to

$$ds^2 = dt^2 \tag{125}$$

and $v = dx/dt = 0$? Although, the speed of the car, as measured by the comoving observer himself $v_{com} \equiv 0$, it is important to realize that, in general $dx \neq 0$ and the metric still is

$$ds^2 = dt^2 - dx^2 = dt^2(1 - p^2) \tag{126}$$

Thus, in general, the quantity $p = dx/dt \neq 0$. And irrespective of the physical meaning of p, Eq.(120), would in this case be modified to $p^2 < 1$ for material particles. In other words, the condition that, for a material particle, one must have $ds^2 > 0$ would tantamount to $p^2 < 1$. But what is the physical significance of p anyway? To see this, let us call the fixed milestone coordinates as X, T, and write

$$X = X(x, t) \tag{127}$$

Now consider the motion of the car as viewed by a given milestone with $X = fixed$, so that $dX = 0$:

$$dX = \dot{X}dt + X'dx = 0 \tag{128}$$

where dot denotes partial differentiation by t and prime denotes differentiation by x. Then, we find that

$$\frac{dx}{dt} = -\frac{\dot{X}}{X'} \tag{129}$$

Using the Lorentz transformation formula:

$$X = \gamma(vt + x) \tag{130}$$

we have $\dot{X} = \gamma v$ and $X' = \gamma$, so that

$$\frac{dx}{dt} = -\frac{\dot{X}}{X'} = -v \tag{131}$$

Now $v = dX/dT$ is the speed of the moving car w.r.t. a fixed milestone. And naturally, $p = dx/dt = -v$ is the *velocity of the fixed milestone with respect to the moving car*. Actually this conclusion could have been drawn without invoking any Lorentz transformation at all. We could have always interpreted p as the velocity of any fixed object w.r.t. the comoving object.

And thus differential of comoving position vector is not at all a meaningless thing and $p = dx/dt$ may be interpreted as the "speed" of the comoving particle w.r.t. a fixed marker, but, in term of comoving coordinates. Note that p is not the speed in comoving frame which is, by definition, zero.

10 Velocity and Speed in GTR

A most general form of spacetime metric is[8,9]

$$ds^2 = g_{00}(dx^0)^2 + g_{\alpha\beta}dx^\alpha dx^\beta + 2g_{0\alpha}dx^0 dx^\alpha \tag{132}$$

Here $\alpha, \beta = 1, 2, 3$ represent the 3 spatial coordinates and 0 represents the temporal coordinate. Note that this general metric can be separated into a spatial and temporal part[8]:

$$ds^2 = d\tau_s^2 - dl^2 \tag{133}$$

where

$$d\tau_s^2 = g_{00}(dx^0 - g_\alpha dx^\alpha)^2; \qquad g_\alpha = -\frac{g_{0\alpha}}{g_{00}} \tag{134}$$

and

$$dl^2 = \left(-g_{\alpha\beta} + \frac{g_{0\alpha}g_{0\beta}}{g_{00}}\right) dx^\alpha dx^\beta \tag{135}$$

Here $d\tau_s$ is the element of *synchronized* proper time i.e., the time interval obtained by considering the

"difference in the values of "time" x^0 for two simultaneous events occurring at infinitely near points."[8]

On the other hand, dl is the element of proper distance.

The arbitrary metric can also be rewritten as[8]

$$ds^2 = d\tau_s^2(1 - v^2) \tag{136}$$

whence the 3-speed v gets defined as

$$v^2 = v^\alpha v_\alpha = \frac{dl^2}{d\tau_s^2} = \left(\frac{dl}{d\tau_s}\right)^2 \tag{137}$$

Here the 3-velocity is defined as

$$v^\alpha = \frac{dx^\alpha}{\sqrt{h}(dx^0 - g_\alpha dx^\alpha)} \tag{138}$$

where

$$h = g_{00} \qquad g_\alpha = -\frac{g_{0\alpha}}{g_{00}} \tag{139}$$

Since $d\tau_s$ is a quantity with direct physical measure and must be real, we see that, the requirement that $ds^2 > 0$ for a material particle tantamounts to the more mundane requirement $v^2 < 1$ as is exactly the case in STR. Conversely since there is no question of $d\tau_s^2$ becoming infinite, occurrence of

$$v^2 = 1 \ implies \ ds^2 = 0 \tag{140}$$

and

$$v^2 > 1 \ implies \ ds^2 < 0 \tag{141}$$

Since this result is independent of any special properties of the coordinates, it must be universally true irrespective of whether there is a supposed "coordinate singularity" or not.

If for a specific case, such as a static metric or the present spherical case (which is a non-stationary metric) $g_{0\alpha} = 0$, we will have

$$d\tau_s^2 = d\tau^2 = g_{00}\, dt^2 \tag{142}$$

and

$$dl^2 = -g_{\alpha\beta} dx^\alpha dx^\beta \tag{143}$$

where $d\tau$ is the usual proper time interval. Further, when all cross terms are zero, i.e., when $g_{\alpha\beta}$ too is diagonal

$$dl^2 = -g_{rr}\, dr^2 \tag{144}$$

Such simplification can occur when the metric is stationary (components do not depend on x^0) and further static. However we may have $g_{0\alpha} = 0$ even in a non-stationary situation (such as a spherically symmetric collapsing fluid). Then

$$v = \frac{\sqrt{-g_{rr}}\, dr}{\sqrt{g_{00}}\, dt} \tag{145}$$

In particular, for a radially symmetric with radial motion,

$$v^R = v = \frac{\sqrt{-g_{rr}}\, dr}{\sqrt{g_{00}}\, dx^0} \tag{146}$$

Corresponding components of 4-velocity $u^i = dx^i/ds$ are

$$u^\alpha = \gamma v^\alpha \tag{147}$$

and

$$u^0 = (\gamma/g_{00}) + \gamma g_\alpha v^\alpha \tag{148}$$

where γ is the usual Lorentz factor. If $g_\alpha = 0$, i.e., the cross term components $g_{0\alpha} = 0$, we obtain.

$$u^0 = \frac{\gamma}{g_{00}} \tag{149}$$

Even for non-radial motion in a spherically symmetric spacetime, we will have,

$$v = \frac{\sqrt{-g_{rr}}\, dr}{\sqrt{g_{00}}\, dx^0} \tag{150}$$

10.1 Speed in Comoving Coordinates

As in the STR case, the entire discussion here implicitly assumed that, the coordinate system used here was not a comoving one. But what would happen if x^α and x^0 were comoving coordinates? For simplification, we will consider the case of spherical symmetry with r and t as comoving coordinates. As we saw in Section 9.1, in such a case Eq.(146) would not at all be meaningless; irrespective of the physical meaning of p^2, in this case, we can demand, instead of v^2, $p^2 < 1$ in order that $ds^2 > 0$. In spherical symmetry, since R is a scalar and is the luminosity distance, the 3-space is interspersed with fixed $R = fixed$ grids which act like fixed "milestones" on the road with respect to which one normally measures the speed of moving particles. But like the moving car, the comoving fluid can measure the speed a $R = fixed$ grid element using his own coordinates r and t. And p will be nothing but the speed of the $R = fixed$ marker w.r.t. the comoving fluid. Alternatively, the speed of the fluid (in terms of comoving coordinates) with respect to the $R = fixed$ marker is $-v$.

11 Nature of the Hilbert Metric at EH

Referring to Eq.(52), we see that as $R \to \alpha_0$, we have

$$\left(\frac{dR}{dT}\right)^2 \to (1 - \alpha_0/R)^2 \tag{151}$$

for any value of E_*. Incidentally Eq.(151) applies to a photon geodesics at any R. This is a direct indication that if indeed $R \to \alpha_0$, kinematically, a material particle would tend to behave like a photon. In any case whenever, Eq.(151) would be valid, whether for a photon at any R or for a material particle at $R \to \alpha_0$, we find that

$$\frac{dR^2}{1 - \alpha_0/R} \to (1 - \alpha_0/R) dT^2 = dz^2 \; (say) \tag{152}$$

Then recalling the radial part of the Hilbert metric, we find that

$$ds^2 = (1 - \alpha_0/R)dT^2 - \frac{dR^2}{1 - \alpha_0/R} \to dz^2 - dz^2 = 0 \tag{153}$$

This behavior could also have been seen by writing

$$ds^2 = (1 - \alpha_0/R)dT^2 \left[1 - (1 - \alpha_0/R)^{-2}(dR/dT)^2\right] \tag{154}$$

and then by using Eq.(151).

In the light of the preceding section, we may promptly understand that $ds^2 \to 0$ at the EH because $v \to 1$ there. Let us see this more explicitly:

Using the general formula (137), it is seen that, for the vacuum Hilbert metric, the 3-speed of a test particle is given by

$$v^2 = \frac{-g_{RR}\, dR^2}{g_{TT}\, dT^2} = (1 - \alpha_0)^2 \frac{dR^2}{dT^2} \tag{155}$$

Now using Eq.(52) in the foregoing Eq., we have

$$v^2 = \frac{E_*^2 - (1 - \alpha_0/R)}{E_*^2} \tag{156}$$

Suppose one would now momentarily heed to the misconception of a mere coordinate singularity at $R = \alpha_0$ and try to evade a direct evaluation of the value of v^2 there. This situation could be something like insisting that the value of $f(\theta) = \sin\theta/\theta$ cannot be evaluated at $\theta = 0$ by directly putting $\theta = 0$ in the expression for $f(\theta)$. Nonetheless one can very well find the value of $f(\theta)$ at $\theta = 0$, by evaluating $f(\theta)$ at $\theta = \epsilon$, where $f(\theta)$ is well defined and then by letting $\epsilon \to 0$. In other words even the pretext of a "coordinate singularity" at the EH cannot prevent the existence of some appropriate 3-speed there. To find this speed we consider the

$$\lim_{R \to \alpha_0} v^2 = 1 \tag{157}$$

And here lies the answer to the physical result - why it takes a "boost" of $v = 1$, to enact a transformation between the free falling coordinate and Hilbert coordinate, at $R = \alpha_0$ and furthermore why it would require a boost of even $v > 1$ for such a coordinate transformation at $R < \alpha_0$; MTW wanted to bypass this question by using an apparently pithy but confusing statement[17] (p. 821):

"The amazing result (a consequence of special algebraic properties of the Schwarzschild geometry, a somewhat analogous to what happens -or, rather, does not happen- to the components of the electromagnetic field, **E** and **B**, when they are both parallel to a boost) is this:..."

They cannot be faulted for this approach because there is no answer to this vital question in the paradigm of finite mass (uncharged) BHs. Further, one must ponder, when GTR says that the choice of coordinate system is arbitrary, why is it not possible to have a $R = fixed$

observer at $R = \alpha_0$ if the EH is really is as regular a region as any other region. The reason is very simple: The ACCELERATION INVARIANT becomes infinite at the EH. In any case, the result $ds^2 = 0$ at $R = \alpha_0$ does not involve the notion of any "fixed observer" and ds^2 is a scalar just like the acceleration invariant.

11.1 Lemaitre Metric

Consider a radial geodesic in the Lemaitre metric comprising comoving radial coordinate r and comoving proper time t for which even the question of any coordinate singularity does not arise. Recalling Section (7.3), see that

$$\frac{dt}{dT} = 1 + \frac{\sqrt{\alpha_0/R}}{1 - \alpha_0/R}\frac{dR}{dT} \tag{158}$$

and

$$\frac{dr}{dT} = 1 + \frac{\sqrt{R/\alpha_0}}{1 - \alpha_0/R}\frac{dR}{dT} \tag{159}$$

so that as $R \to \alpha_0$ or $R/\alpha_0 \to 1$, we will have

$$\frac{dt}{dT} = \frac{dr}{dT}; \qquad R \to \alpha_0 \tag{160}$$

Or,

$$dt = dr; \qquad R \to \alpha_0 \tag{161}$$

Therefore, for a radial geodesic, in this limit, we find

$$ds^2 = dt^2 - dt^2 = 0 \tag{162}$$

Note that we do not require here the existence of any fixed observer and the coordinates used are comoving coordinates. This result directly shows that the speed of the free falling particle at the EH, *as seen by any other observer* (and not obviously by the comoving observer himself, see Sec. 9, 10, and 12), is $v = 1$ in violation of the principles of relativity. This latter result can be independently found by noting that

$$v^2 = \frac{\alpha_0}{R}\left(\frac{dr}{dt}\right)^2 \tag{163}$$

and since, $dr = dt$ at $R = \alpha_0$, $v^2 = 1$ Thus if there would be any spacetime beneath the EH where the free falling observer might enter, because of focusing action due to gravitational field, the speed of the observer would **exceed** the speed of light.

As to the occurrence of $ds^2 = 0$ at the EH, since ds^2 is an invariant, this anomalous behavior must be present in any coordinate system and there is just no question of this being removable. Hence the EH must be a true physical singularity. In other words, the EH itself must be the central singularity, the location of the mathematical massenpunkt. This is possible when $R_g = 2M_0 = 0$, or $M_0 = 0$. And the comoving proper time to reach the EH or the central singularity would be infinite, indicating that the free falling observer never reaches the EH in exact agreement with the conclusion made by the distant observer ($T_H = \infty$).

11.2 Speed in Finkelstein Coordinate

The above result, in view of its importance, is worthwhile another rederivation without referring at all to the underlying Hilbert metric. The following powerful exercise to this effect has been carried out by Robertson[45]:

In the stationary field, as mentioned before, energy per unit rest mass of the test particle is conserved:

$$E_* = p_0/m_0 = g_{0i}(dx^i/ds) = qg_{00} + pg_{0R} = constant \tag{164}$$

where

$$q \equiv dT_*/ds; \qquad p \equiv dR/ds \tag{165}$$

The Christoffel symbols for the radial part of the metric are worked out to be

$$\Gamma^0_{00} = \mp\alpha_0/R^3,\ \Gamma^0_{10} = \Gamma^0_{01} = \alpha_0/2R^2(1+\alpha_0/R) \tag{166}$$

$$\Gamma^0_{11} = \mp\alpha_0/R^2(1+\alpha_0/2R) \tag{167}$$

$$\Gamma^1_{11} = \mp\alpha_0/R^2(1+\alpha_0/R) \tag{168}$$

$$\Gamma^1_{10} = \Gamma^1_{01} = \pm\alpha_0/R^3 \tag{169}$$

$$\Gamma^1_{00} = \alpha_0/R^2(1-\alpha_0/R) \tag{170}$$

By using these Christoffel symbols, one can form the geodesic equations and the first equation is a statement of the energy momentum conservation :

$$g_{ij}p^ip^j = m_0^2 \tag{171}$$

and which yields

$$(1-\alpha_0/R)q^2 \pm 2\frac{\alpha_0 pq}{R} - (1+\alpha_0/R)p^2 = 1 \tag{172}$$

This result however also follows by just dividing both sides of the metric (94) by ds^2. The other two geodesic equations, on the other hand, are obtained after lengthy algebra as

$$(dq/ds) \mp \frac{\alpha_0^2 q^2}{2R} + \alpha_0(1+\alpha_0/R)\frac{pq}{R^2} \mp \alpha_0(1+\alpha_0/2R)\frac{p^2}{R^2} = 0 \tag{173}$$

and

$$dp/ds + \alpha_0(1-\alpha_0/R)\frac{q^2}{2R^2} \pm \alpha_0^2\frac{pq}{R^3} - \alpha_0(1+\alpha_0/R)\frac{p^2}{2R^2} = 0 \tag{174}$$

Eq.(172) can be compactified as

$$q^2 - p^2 - (\alpha_0/R)(q-p)^2 = 1 \tag{175}$$

and Eqs.(172) and (174) can be combined to obtain

$$\frac{dp}{ds} + \frac{M_0}{R^2} = 0 \tag{176}$$

which is just the Newtonian free fall equation with time as the proper time. If the free fall starts at $R = R_i$, by using the fact $dp/ds = pdp/dR$, the above equation can be integrated to yield

$$p = (dR/ds) = -\sqrt{\alpha_0/R - \alpha_0/R_i} \tag{177}$$

Under the assumption of $\alpha_0 > 0$, this equation would show that the test particle would reach the EH in a finite proper time and can apparently cross the same. But let us examine the case more closely by keeping α_0 as a truly unknown integration constant.

Subtracting Eq.(174) from Eq.(173), we have,

$$\frac{d(q-p)}{ds} - M_0\frac{(q-p)^2}{R^2} = 0 \tag{178}$$

Or,

$$\frac{d(q-p)}{dR} = -\frac{M_0}{R^2} = \frac{dp}{dR} \tag{179}$$

This can be integrated to yield

$$q = p + \frac{1}{C_1 + p} \tag{180}$$

where C_1 is an integration constant. From. Eq.(177), we see that, $p = 0$ at $R = R_i$, so that $C_1 = 1/q_i$. Then considering the value of q_i from Eq.(167), we obtain

$$C_1 = \sqrt{1 - \alpha_0/R_i} \tag{181}$$

Therefore, finally, we have,

$$q = p + \frac{1}{\sqrt{1 - \alpha_0/R_i} + p} \tag{182}$$

Again from Eq.(177), we find that, at $R = \alpha_0$, we have

$$p = p_H = -\sqrt{1 - \alpha_0/R_i} \tag{183}$$

so that at $R = \alpha_0$ one has

$$q = q_H = p_H + \frac{1}{\sqrt{1 - \alpha_0/R} - \sqrt{1 - \alpha_0/R}} = \infty \tag{184}$$

The fact that $q_H = dT_*/ds = \infty$ directly shows that *the originally timelike radial geodesic would become null, $ds = 0$, if there would be any EH*. This demands that

(i) The EH cannot be a mere coordinate singularity and

(ii) EH must not ever form, i.e., its proper time of formation must be infinite which again follows from Eq.(73) by realizing that $ds^2 = 0$ at $R = \alpha_0$ implies $\alpha_0 = 0$.

Further, again using the quadratic Eq.(172), it can be found that, for $\alpha_0/R = 1$, we have

$$p_H = \frac{\pm q_H \mp \sqrt{q_H^2 - 2}}{2} = \frac{\pm q_H \mp q_H}{2} = 0; \qquad because\ q_H = \infty \tag{185}$$

The occurrence of $q_H = dR/ds = 0$ shows that $dR = 0$ at the EH, i.e., $R = R_g$ is one of the extrema of R. But R has only two extrema; one at $R = R_i$ and another at $R = 0$. Thus we must have $R = R_g = \alpha_0 = 2M_0 = 0$. Mathematically occurrence of $dR = 0$ at $R = \alpha_0$ indicates that the EH is the endpoint beyond which the affine parameters cannot be extended.

One can also formally find the 3-speed of the test particle in Finkelstein metric by using Eq.(129):

$$v^2 = \frac{(g_{0R}^2 - g_{RR}g_{00})v^R v^R}{(g_{00} + g_{0R}v^R)^2} \tag{186}$$

where $v^R = dR/dT_*$. Since

$$g_{00} \to 0;\ g_{0R} \to \pm 1; as \qquad R \to \alpha_0 \tag{187}$$

once more, we get the fundamental result $v^2(EH) = 1$.

We find now that the result $v^2 = 1$ at EH follows in several other coordinates too. Thus while the values of v^2 are different in different coordinates for $R > \alpha_0$, one must have $v^2 = 1$ at $R = \alpha_0$. And hence if there would be any spacetime below the EH, one would have $v^2 > 1$ for a material particle in direct violation of relativity. And therefore no gravitational collapse must be able to produce any true BH. To explicitly see this we now turn to some hitherto unknown basic features of arbitrary spherical gravitational collapse.

12 Spherical Gravitational Collapse

When a self-gravitating fluid undergoes gravitational contraction, by virtue of Virial Theorem, part of the gravitational energy must be radiated out. Thus the total mass energy, M, $(c = 1)$ of a body decreases as its radius R decreases. But in Newtonian regime $(2M/R \ll 1, G = 1)$, M is almost fixed and the evolution of the ratio, $2M/R$, is practically dictated entirely by R. If it is *assumed* that even in the extreme general relativistic case $2M/R$ would behave in the *same Newtonian* manner, then for sufficiently small R, it would be possible to have $2M/R > 1$, i.e., trapped surfaces would form[46].

Unfortunately, even when we use General Relativity (GR), our intuition is often governed by Newtonian concepts, and thus, intuitively, it appears that, as a fluid would collapse, its gravitational mass would remain more or less constant so that for continued collapse, sooner or later, one would have $2M/R > 1$, i.e., a "trapped surface" must form. The singularity theorems thus start with the *assumption* of formation of trapped surfaces[46]. However, many readers (from experimental astronomy, particle physics, quantum-gravity and quantum information fields) may not be even aware that the *occurrence of trapped surface is a conjecture* and the Singularity Theorems are based on this conjecture. Further, in the literature, there have been attempts to find out "sufficient and necessary condition" for realization of this conjecture of trapped surfaces for spherical collapse[47] without showing that such "necessary and sufficient conditions" *are fulfilled*. Nonetheless, we have found that, by superficially going through such papers, many readers (erroneously) think that the conjecture of trapped surface has been proved. The actual situation regarding this has been succinctly brought out in the following statement by Dadhich[47]:

"it is necessary to point out that a demonstration of the trapped surface conjecture remains elusive."

Thus in view of the fact that the formation of trapped surfaces is an open question even in 2005, by assuming formation of the same, Hawking, Penrose and others virtually assumed what they, at least ought to have attempted to prove. It is believed that the pressure free O-S dust collapse showed the formation of trapped surfaces. But in Section(6), we have shown that actually it is not so. Even if one would accept for a moment that in the O-S problem, the mass of the dust ball is finite in violation of thermodynamics and hydrodynamics, the mental pictures obtained from this physically absurd case cannot be generalized to all physical cases. Yet, after discussing dust collapse, MTW write (p. 984):

"Non-free-fall collapse is similar to the collapse depicted here" (!) while, in the context of the assumptions behind the singularity theorems, they too admit that (p. 935):

" All these conditions, except the trapped surface, seem eminently reasonable for any physically realistic spacetime!"

In fact, there are specific examples that trapped surfaces do not form. For example, in the cosmological context, Nariai Metric[48] is the first specific example of non-occurrence of trapped surfaces. In this context, Dadhich's comment about this "assumption" of trapped surfaces is[47]:

"It is noteworthy that violation of this assumption entailed no unphysical features. This assumption seriously compromises, as is demonstrated by this example, the generality of the theorems (i.e., singularity theorems, author). It has always been looked upon as suspect."

Another work on spherical collapse using premeditated specific metric finds that collapse would not produce any Horizon because of heat flow, i.e., because of decrease of M during collapse[50]. This specific example focused attention only at the boundary of the fluid and had it treated the inner mass shells, it might have found absence of trapped surfaces too.

In 1990, Senovilla constructed a specific model of a cylindrically symmetric universe without any trapped surface[51]. In 2002, Goncalves, in a more general manner, showed the absence of trapped surface and singularities in cylindrical collapse[52].

In the cosmological context, M may not be properly defined at all, and in case, it would be defined, it may remain fixed because radiation cannot leave the universe. But for isolated bodies, M necessarily decreases due to emission of radiation, and thus, it is more likely that trapped surfaces do not form for isolated bodies.

In any case, once the formation of a "trapped surface" is assumed, then, a set of very reasonable assumptions would show that the collapse would become irreversible. Physically, once trapped surface formation is *assumed*, it would appear that the sign of the pressure gradient force would reverse and thus pressure would aid the collapse rather than hinder the same. Furthermore, since outgoing radiation too would turn inward, no mass energy would escape and the mass energy enclosed within a shell $M(r,t)$, where r is a comoving coordinate, would not decrease any longer. Hence if the collapse would lead to a BH with an all encompassing Event Horizon (EH), it would naturally appear that the mass of the BHs must be finite. Thus the intuitive and *apparently* correct notion of finite mass BHs rests on the *assumption* of formation of trapped surfaces, which, in turn, rests on our intuitive (but incorrect) idea that even when gravity would be so strong as to trap even light M would remain more or less unchanged *as in the Newtonian case*.

However, the following **exact** treatment would unequivocally show that, trapped surfaces do not form in any spherical collapse[53]:

If we take the signature of spacetime as (+,-,-,-) the spherically symmetrical metric for an arbitrary fluid, in terms of comoving coordinates t and r is[54-56]

$$ds^2 = g_{00}dt^2 + g_{rr}dr^2 - R^2(d\theta^2 + \sin^2\theta d\phi^2) \tag{188}$$

where $R = R(r,t)$ is the Invariant Circumference coordinate and happens to be a scalar. Further, for radial motion with $d\theta = d\phi = 0$, the metric becomes

$$ds^2 = g_{00}dt^2(1 - x^2) \tag{189}$$

whence

$$(1 - x^2) = \frac{1}{g_{00}}\frac{ds^2}{dt^2} \tag{190}$$

and the auxiliary parameter

$$x = \frac{\sqrt{-g_{rr}}}{\sqrt{g_{00}}}\frac{dr}{dt} \tag{191}$$

For a fluid element at $r = fixed$, obviously $x = 0$ because $dr = 0$. However, dr, in general, is not zero; otherwise the metric in comoving coordinates would be

$$ds^2 = g_{00}dt^2 - R^2(d\theta^2 + \sin^2\theta d\phi^2) \tag{192}$$

and which is *not* the case. The comoving observer at $r = r$ is free to do measurements of not only the fluid element at $r = r$ but also of other objects: If the comoving observer is compared with a static floating boat in a flowing river, the boat can monitor the motion of other boats or the pebbles fixed on the river bed. Here the fixed markers on the river bed are like the background $R = constant$ markers against which the river flows (recall here Sections 9.1 and 10.1). If we intend to find the parameter x for such a $R = constant$ marker, i.e., for a pebble lying on the river bed at a *fixed* R, we will have[53]

$$dR(r,t) = 0 = \dot{R}dt + R'dr \tag{193}$$

where an overdot denotes a partial derivative w.r.t. t and a prime denotes a partial derivative w.r.t. r. Therefore for the $R = constant$ marker, we find that

$$\frac{dr}{dt} = -\frac{\dot{R}}{R'} \tag{194}$$

and the corresponding x for the $R = constant$ marker is

$$x = x_c = \frac{\sqrt{-g_{rr}}\, dr}{\sqrt{g_{00}}\, dt} = -\frac{\sqrt{-g_{rr}}\, \dot{R}}{\sqrt{g_{00}}\, R'} \tag{195}$$

Using Eq.(189), we also have

$$(1 - x_c^2) = \frac{1}{g_{00}} \frac{ds^2}{dt^2} \tag{196}$$

Now let us define[54-56]

$$\Gamma = \frac{R'}{\sqrt{-g_{rr}}} \tag{197}$$

and

$$U = \frac{\dot{R}}{\sqrt{g_{00}}} \tag{198}$$

so that the combined Eqs. (194), (196) and (197) yield

$$x_c = -\frac{U}{\Gamma}; \qquad U = -x_c \Gamma \tag{199}$$

As is well known, the gravitational mass of the collapsing (or expanding) fluid is defined through the equation[17, 54-56]

$$\Gamma^2 = 1 + U^2 - \frac{2M(r,t)}{R} \tag{200}$$

Then the two foregoing equations can be combined and transposed to obtain

$$\Gamma^2(1 - x_c^2) = 1 - \frac{2M(r,t)}{R} \tag{201}$$

By using Eqs.(198) and (199) in the foregoing Eq., we have

$$\frac{R'^2}{-g_{rr} g_{00}} \frac{ds^2}{dt^2} = 1 - \frac{2M(r,t)}{R} \tag{202}$$

Recall that, the determinant of the metric tensor is always negative[11]:

$$g = R^4 \sin^2\theta\, g_{00}\, g_{rr} \leq 0 \tag{203}$$

and when we remove $R^4 \sin^2\theta$ factor, the determinant must be non-zero at a non-singular region

$$g_{rr}\, g_{00} < 0; \qquad -g_{rr}\, g_{00} > 0 \tag{204}$$

Further for the signature chosen here, at a non-singular region, $ds^2 > 0$ for all material particles. Then noting Eq.(204), it follows that the LHS of Eq.(202) is *always positive*. So must then be the RHS of the same Eq. which implies that

$$\frac{2M(r,t)}{R} < 1 \tag{205}$$

This shows, in the utmost general fashion, that *trapped surfaces are not formed in spherical collapse or expansion* of isolated bodies. And if we include motion of photons and the mathematical possibility of material particles hitting the singularity, we will have $ds^2 \geq 0$, and accordingly,

$$\frac{2M(r,t)}{R} \leq 1 \tag{206}$$

This result was first derived by Mitra[2,3,57,58,59]. However those papers did not mention that the result

$$\frac{U}{\Gamma} = \frac{\sqrt{-g_{rr}}\, dr}{\sqrt{g_{00}}\, dt} = -\frac{\sqrt{-g_{rr}}\, \dot{R}}{\sqrt{g_{00}}\, R'} \tag{207}$$

was due to the fact that for a $R = fixed$ marker $dR = \dot{R}dt + R'dr = 0$, and it is only for a such a fixed marker, $|\dot{R}/R'| = |dr/dt|$. As a result, somewhat justifiably, some of the readers like J. Ehlers and S. Senovilla thought that, in Eqs.(194) and (199), the partial derivatives of R were being confused as total derivatives (Private communication, 2000). But we clearly see now that it is simply the result of watching a $R = fixed$ marker. Irrespective of the debate, whether there can be $R = fixed$ observer at the EH or not, there is no denying of the fact that, in spherically symmetric spacetime is enumerated by $R = fixed$ markers. For instance the EH is marked by $R = \alpha_0 = fixed$ marker and the central singularity is marked by $R = 0 = fixed$ marker (where there may not be any fixed observer).

Note that to obtain this fundamental result of non-occurrence of trapped surfaces in arbitrary spherical collapse, we did not require to invoke the notion of any 3-speed. However having done so, we can easily find that trapped surfaces do not form in order that the speed of the collapsing fluid (as seen by any observer) remains subluminal.

Going back to Eqs.(190) and (191), we quickly identify x as the 3-speed of an object (not necessarily fluid element) measured by the comoving observer at $r = r$. Obviously, the 3 speed of the fluid element itself at $r = r$ is $x = 0$. But here x_c is the 3-speed of the $R = constant$ marker, i.e., the pebble fixed on the river bed, and is non-zero.

12.1 3-speed of the Fluid

Since x_c is the speed of the $R = constant$ marker w.r.t. the $r = constant$ observer, the speed of the $r = constant$ observer, i,e, *fluid itself*, w.r.t. the $R = constant$ marker is the *negative* of x_c: i.e.,

$$v = -x_c = -\frac{\sqrt{-g_{rr}}\, dr}{\sqrt{g_{00}}\, dt} = +\frac{\sqrt{-g_{rr}}\, \dot{R}}{\sqrt{g_{00}}\, R'} \tag{208}$$

In terms of $v = -x_c$, let us rewrite Eqs. (199) and (201) as

$$U = +v\Gamma \tag{209}$$

and

$$\Gamma^2(1 - v^2) = 1 - \frac{2M(r,t)}{R} \tag{210}$$

When these results were first presented[2], unlike the present case, the physical meaning of v, i.e., it is the speed of the fluid w.r.t. a neighbouring $R = constant$ marker was not clearly shown. Though it was emphasized that v was the speed of the fluid involving comoving coordinates rather than w.r.t. the same, many readers could not grasp the difference. They mistook v for the latter, and since $v_{com} \equiv 0$, they had difficulty in appreciating Eq.(208).

Now using Eq.(206) in (210), we find that both sides of it are positive and hence

$$v^2 \leq 1 \tag{211}$$

Thus, in retrospect, we see that non-occurrence of trapped surface is a direct consequence of this *cornerstone of relativity*: $v^2 \leq 1$ for material particles and photons. If γ is the corresponding Lorentz factor, Eq. (210) of general relativistic fluid motion acquires the appealing form:

$$\frac{\Gamma^2}{\gamma^2} = 1 - \frac{2M(r,t)}{R} \tag{212}$$

This is the *Master Equation* of spherical GR GC. In the absence of any trapped surface, a collapsing fluid will always keep on radiating and $M(r,t)$ would keep on decreasing. In case it would be *assumed* that, the collapse would continue all the way up to $R = 0$, then the constraint (205) demands that $M(R_0 = 0) = 0$ too. And this is the reason that all BHs (even if they would be assumed to exist) must have $M = 0$. This is exactly the result which follows from physical interpretation of the most important *exact* solution of GTR, namely, the Hilbert solution; $\alpha_0 \equiv 0$. If the collapse would not proceed up to $R \to 0$, then of course there would be finite radius and finite mass objects with $\alpha = 2M > 0$; $R_0 > 2M$. But such objects could be extremely hot in which photons and neutrinos will remain virtually trapped. However, in a very strict sense, these objects would still be collapsing on a very long secular time scale and the condition $z < 2$ would not be applicable to them.

13 How do others get BH?

A question that would naturally arise here is how do other authors regularly obtain BHs or other singularities such as "Naked Singularities" either by numerical means or analytical means?

Physically, formation of an EH or a NS can be considered as the formation of a surface with gravitational redshift

$$z = (1 - 2M/R)^{-1/2} - 1 = \infty \tag{213}$$

In comparison, cold neutron stars have $z \sim 0.1 - 0.2$.

Recall that when faithful attempts are made, even now, many numerical physicists admit that they are still unable to simulate the "bounce" in supernova implosion (range of $z = 0.1 - 0.2$)[60]. On the other hand many eager numerical physicists routinely claim to find BH ($z = \infty$) or Naked Singularity formulation in their studies! The main frame computers came into being approximately 50 years back (after the death of Einstein in 1955) and the codes for simulating H-bomb were getting ready thereafter. Essentially the same codes were applied to study gravitational collapse and despite having a computing resource probably thousand times less that present day super computers, it was declared that "BH formation" has been found numerically[54,55]! J. Wheeler got excited over this numerical result and brought it to the attention of the world in a meeting of the "American Association for the Advancement of Science" in 1967. Alas, many physicists are still struggling with the much simpler problem of simulating the supernova bounce[60]. One may also wonder, if the evidence for BH formation was already there in 1967, why should many authors be still working on the same problem and reporting formation of BHs?

Most likely numerically, one may be actually exploring whether the collapse process suddenly halts with such a modest range of $z \sim 0.1$. And in the high z regime, one must necessarily consider the entire physics of GC including the radiative transport part in an exact way. However in practice it is not possible to do so; nonetheless, a given paper must come out with "a result". So if there is no indication of the collapse coming to a sudden halt (at a modest value of z), numerical physicists may be declaring that "a BH has formed"! Most of the complex physical problems must be solved numerically and numerical analysis has its own limitations. And it is necessary that a numerical analysis makes various simplifications and assumptions because otherwise the problem may not be tractable at all. Part of the inevitable hidden assumptions and simplifications in a given "n'th paper" is primarily known when the next paper in the series is written : "..the present work is an improvement over the previous one because of ...".

While such an approach is very much justified for most of the complex numerical problems it is not appropriate for the unique problem of GC involving $K \to \infty$, $z \to \infty$, $\rho \to \infty$, $p \to \infty$, temperature $T \to \infty$ and probably $q \to \infty$ (though $L \sim 4\pi R^2 q \to 0$).

It is the misinterpretation of the OS result of BH formation (which actually corresponds to $p = 0, \rho = 0, M = 0$) and Wheeler's announcement of having numerically found the evidence of BH formation, which laid the foundation of modern BH and GC research; though, which, as we have found, has moved on completely wrong tracks. One wonders, if in the history of science, whether such a prolonged regressive research has ever happened.

While one can legitimately question the accuracy and realm of numerical simulations, one may ask "what about supposed analytical treatments which regularly claim occurrence of either BHs or Naked Singularities?" To be specific we consider some of the very recent analytical works on this topic :

1. Gravitational Collapse from a smooth initial data[61]
A. Mahajan, R. Goswami & P.S. Joshi (gr-qc/0312015)
2. Gravitational collapse of an isentropic perfect fluid with a linear equation of state[62]
R. Goswami & P.S. Joshi (gr-qc/0406052)
3. Black hole formation in perfect fluid collapse[63]
R. Goswami & P.S. Joshi (gr-qc/0310122).

These works in a sense represent mainstream theoretical research on GC

The basic reason that these authors claim to find the occurrence of spacetime singularities even when they consider a fluid with finite pressure, unlike the OS case, is that they believe in the occurrence of "Apparent Horizon", i.e., formation of the $R(r,t) = 2M(r,t)$ surface at a finite value of R. And why do they believe so?

They believe in/find "Apparent Horizon" and trapped surface formation because they **ignore the crucial fact that there cannot be any GC without accompanying emission of radiation by the H- K process**. Like many other authors working on GC, they too treat the GC as an applied mathematics problem by throwing out the crucial physics part of H-K process. In other words they ignore the fact, if there would be pressure and internal energy and unless the fluid is a pure incoherent photon gas with an *exact* adiabatic index of $\gamma = 4/3$ and $E_* = \infty$, the mass function must keep on decreasing as the fluid would keep on radiating. To see this let us consider a crucial GR equation for "adiabatic" collapse[56]:

$$\frac{1}{\sqrt{g_{00}}}\dot{M}(r,t) = -4\pi R^2 pU \tag{214}$$

where $U = R'/\sqrt{g_{00}}$. Here the mass-energy content within a given shell is

$$M(r,t) = \int_0^r 4\pi R^2 \rho R' \, dr \tag{215}$$

For a static case, one has, $R(r,t) = R(r)$ and $R' = dR/dr$. In such a case, one recovers the more familiar formula

$$M(r) = \int_0^R 4\pi R^2 \rho \, dR \tag{216}$$

Let us focus attention on a given mass-shell, i.e., one having $r = fixed$ (the way $M(r,t)$ is defined in Eq.[215]). Since $r = fixed$ for this surface,

$$\frac{\partial M}{\partial t} = \frac{dM}{dt} \tag{217}$$

Then from Eq.(214), we have

$$\frac{dM}{dt} = -4\pi R^2 \, p \, \dot{R} \tag{218}$$

For adiabatic collapse, this condition is obviously satisfied for the boundary of the fluid, where $p = 0$ and $M = constant$. But in the absence of any heat transport, the mass-energy content of arbitrary shell must be constant, i.e., one must have $dM(r,t)/dt = 0$ everywhere for an adiabatic collapse *without any heat transport.*

Thus for adiabatic collapse, we must have

$$4\pi\, R^2\, p\, \dot{R} = 0 \tag{219}$$

So if one would assume adiabatic collapse (or expansion), i.e., $\dot{R} \neq 0$*, then one must have* $p = 0$*. Thus all adiabatic collapse cases in GR actually correspond to pressureless dust collapse even if one has an elaborate formalism involving a perfect fluid, scalar field, or anything else*. Incidentally, Eq.(219) also shows that a singular state with $R_0 = 0$ is necessarily non-radiative, and this may be telling that, there cannot be any (global) naked singularity if the problem would be treated realistically. This could be specifically seen at least in one paper on radiative collapse:

When the heat transport vector is set to zero, not only the radiation energy density $q = 0$, but the fluid pressure too becomes so, $p = 0$[64].

Further, we have seen in Section 6, that, even if one assumes an inhomogeneous dust (which may give rise to apparent Naked Singularities), even mathematically, there can be only a homogeneous dust. Thus all adiabatic collapse studies actually correspond to the OS collapse which in turn corresponds to formation of a $M = 0$ BH.

And if one insists or pretends that the fluid has finite pressure and internal energy and yet undergoing "adiabatic" collapse, Eq. would tell that in such a case, $\dot{R} = 0$, i.e., *physically there is no collapse at all* although, as an empty applied mathematics exercise, one would feel that he/she is studying GC!

Sometimes these authors explicitly assume a constant mass function $F(r,t)$, ($M(r,t)$ in our notation)

$$F(r,t) = r^3 \mathcal{M}(r) \tag{220}$$

Sometimes they try to convey an impression that they are explicitly considering the temporal evolution of the mass function by splitting it in terms a function $Q(t)$. But, in the absence of explicit incorporation of the H-K process, this temporal variation exists only as an empty applied mathematics exercise.

Once the assumption of formation of apparent horizon ($R(r,t) = 2M(r,t) > 0$) is made, there would be a fictitious singularity (finite τ_c). Then depending on other peripheral details, such as density profile, initial data, simplifications about assumed EOS, naturally, one would find either a BH or a Naked Singularity. In particular, if in the given applied mathematics problem, the neighborhood of $r = 0$ would get trapped earlier than the singularity, then it would be covered (BH) otherwise, it would be "Naked".

What would be the correct interpretation of such works? When we treat the problem as a genuine physics problem, as discussed in the introduction, gravitational contraction would not yield any radiation only if the pressure and internal energy of the fluid are zero. And if we do not trample physics in favor of workable applied mathematics, pressure and internal energy can be zero only when density is zero which in turn demands $M = 0$. Thus if we honor physics, the mass function considered in the above three papers is actually

$F(r, t) = F(r) = 0$ and hence there is no question of any apparent horizon or trapped surface formation at a finite R. And on the other hand, apparent horizon can only form at $R = 0$ indicating the formation of a zero mass BH. However, the very concept of BHs implicitly assume $M > 0$, and therefore some authors call such zero mass BHs as "marginally naked singularities"[65].

Although there could be innumerable examples of "research" on GR GC, to highlight the average state of the subject we may mention of a recent "review paper" entitled "Gravitational Collapse and Naked Singularities" by T. Harada (gr-qc/0407109)[66]. This paper does not contain a single numbered equation and it hardly discusses any physics associated with any real matter having flesh and blood, i.e., H-K mechanism and radiation/heat transport. The absence of such physics is overcompensated by several "spacetime diagrams" none of which is based on any *exact solution* involving any *real matter*. At one place, Harada mentions of "some appropriate time and radial coordinate r and t". However, these "appropriate" coordinates are never used directly. Instead they are baptized into modern GC research by means of the following complex transformations: $x = \ln(-r/t)$ and $\tau = \ln(-t)$. And it is never explained why it is necessary to have complex x and t to describe a physical problem. This example illustrates the status of a large chunk of modern GC research:

• It is something like an inverted pyramid where various blocks of the pyramid are bonded by mental pictures (spacetime diagrams) rather than by physical or analytical solutions. The apex of the pyramid is believed to be lying on the solid foundation of the OS solution which in turn is believed to have shown the formation of trapped surfaces. But we know that this is only a chimera; a result of assuming attainment of a $\rho = \infty$ state even when pressure is assumed to be fixed at $p = 0$. In reality the apex of the pyramid hangs in the vacuum of $\rho = 0$ and not on the turf of $\rho = finite$.

In fact Joshi, Goswami, Mahajan, Harada and many GC authors *do not consider any radiation transport at all*. Such exercises, as if, are something like making castles in the vacuum and then debating about the density profile or color or shape of the castles. However, in the purely Newtonian regime of extremely weak gravity, $z \to 0$, $K \to 0$, when the H-K process hardly alters the mass-energy density, non-radiative collapse may yield an approximately correct physical picture. But to study such Newtonian collapse, one need not use GR GC at all and derive conclusions meant for $z \to \infty$, $K \to \infty$.

Fortunately many GR workers, particularly, those works originating from Brazil and South Africa, more often than not include radiation transport in GR GC. This does not however mean that if any numerical treatment would consider the radiation transport problem in an approximate or premeditated form, it would give the correct physical result. Because, to start with, one must not assume the formation of any apparent horizon/EH at a finite radius and the code or the analytical treatment must be able to accurately handle the problem even if one would approach $z = \infty$, $K = \infty$, $\rho = \infty$, $p = \infty$, $T = \infty$ and $q = \infty$ too. On the other hand, if some numerical treatment would attempt to find the result without any premeditated idea it might find that GR GC gives rise to ECO rather than a BH. Very significantly, the first such computation has already been carried out by Mosquera Cuesta, Salim and Santos[67];

"Here we address the issue using a general relativistic perturbation theory assessment where dissipation through heat flow produces a null radiation. It is shown that a purely gravitational eternally collapsing object (GECO) can indeed be formed only during the collapse of a supermassive star, whose radius after collapse is larger than the Schwarzschild's"

The handling of EOS and associated radiation transport is much simpler for a low density supermassive star in comparison to a stellar mass MECO having supra-nuclear density. But we believe that even in the latter case, one would find formation of ECOs provided the effect of radiation trapping is properly incorporated.

13.1 A crucial point on inclusion of radiation pressure

To appreciate this, one has to recall the definition of "Gravitational Redshift" z associated with any spherical object of mass M and radius

$$1 + z = (1 - 2M/R)^{-1/2} \tag{221}$$

For the Neutron Stars, supposed to be highly compact, the value of z lies in the modest range of $0.1 - 0.2$. But for the EH of a Schwarzschild BH, one has $z = \infty$. If one plots the compact objects against z, there would be and *infinite gap* between a NS and a BH! And it believed that during the collapse, once the star crosses the limit of $z = 0.1 - 0.3$ (NS), then it directly jumps to $z = \infty$ (BH). Obviously the star has to pass through the range of say $z = 1$, $z = 10$ and so on before arriving at $z = \infty$. In particular, once, the collapse crosses the limit, $(1 + z) > \sqrt{3}$, then, all the photons/neutrinos originating at a given point cannot escape, and some of them stay put with the emission region. The escape angle ψ of even photons/ neutrinos (not to talk of matter) becomes extremely narrow for $z > 2.0$ even in vacuum[68,69]:

$$\sin\psi < \frac{3\sqrt{3}GM}{Rc^2}(1+z)^{-1} \approx 2.5(1+z)^{-1}; \quad z \gg 1 \tag{222}$$

As a result, once the collapse is strong enough, for sufficiently massive stars, the radiation generated during the collapse *becomes virtually trapped* within the collapsing body as $\psi \to 0$ (almost). The escape probability depends on the escape solid angle $\sim \psi^2 \sim (1+z)^{-2}$ for small ψ. Consequently, the trapped radiation pressure increases as $\sim (1 + z)^2$ for $z \gg 1$. Thus in order to reach the $z = \infty$ BH stage, one must pass over the stage when trapped radiation pressure would start rising dramatically. Sooner or later, the luminosity of the trapped radiation must attain the appropriate local Eddington value (*because it is very much a finite quantity*) at appropriate high value of z. And it is this *trapped radiation and the associated pressure which virtually halt the collapse* to make a quasi steady ECO[68, 69]

Thus, almost, by definition, **any numerical computation which would successfully incorporate, the fundamental GR effect of bending of radiation in strong gravitation** (i.e., large z), would find an ECO much before it would find any BH!

And when would the BH be formed? Note that the mass-energy is defined the energy content of the body w.r.t. a distant observer in an asymptotic flat spacetime (S_∞). Thus though the local value of the Eddington luminosity is

$$L_{ed} = 1.3 \times (1+z)\, 10^{38}\, (M/M_\odot)\; erg/s \tag{223}$$

its value as seen by S_∞ is

$$L_{ed,\infty} = \frac{L_{ed}}{(1+z)^2} = \frac{1.3\times\; 10^{38}\,(M/M_\odot)\; erg/s}{1+z} \tag{224}$$

Then the time scale of the GR H-K process for the ECO radiating at its Eddington limit is

$$T_\infty \sim \frac{Mc^2}{L_{ed,\infty}} = \frac{2\times 10^{54}\, erg/s\,(M/M_\odot)}{L_{ed,\infty}} \sim (1+z)\, 10^{16}\; s \tag{225}$$

Thus astrophysically, we would see the ECO life time to be more than Hubble time even for $z \sim 10^3$[32].

And as the BH stage would be, mathematically, arrived at, i.e., as $z \to \infty$, one would have $T_\infty \to \infty$ too. During this infinite journey, the MECO would radiatiate out its entire mass energy to find peace as a $M = 0$ BH.

Another academic duplicity in the GC works claiming formation of EH is the following: The outer boundary of the collapsing fluid is well marked by a $R = R_0(t)$ marker which is always considered as a good spacelike coordinate. Even if the fluid would be assumed to collapse inside its own EH, $R = R_0$ is still considered as a good coordinate. Then what about the refrain : "R breaks down as a radial coordinate inside the EH"? In fact, Joshi, Goswami & Mahajan mention R as the "Physical Radius" (when actually the physical or proper radius is l). If any case, if $R = R_0$ is the "physical radius" (actually externally perceived invariant radius), then how can $R = R_0$ be simultaneously considered as a "bad coordinate" within the EH?

Is there any possibility that any author would ever analytically show the formation of either a finite mass BH or a naked singularity (for collapse of isolated bodies)? To achieve this, one has to first prove (i) the formation of trapped surfaces in a general case rather than only discuss conditions for formation of trapped surfaces. Before one would do this, one has to show that our derivation of non-occurrence of trapped surfaces or our result that for an initially timelike radial geodesic $ds^2 \to 0$ at the EH are incorrect. And the fact is that no body has been able to so.

On the other hand, the result that there cannot be either any EH or any apparent horizon in order that the timelike radial worldlines of an infalling test particle or an element of a collapsing fluid does not become null has been independently verified by Leiter and Robertson[32].

This result that spherical gravitational collapse of sufficiently massive bodies try to asymptotically approach the zero mass BH state is sometimes interpreted as the occurrence of Naked Singularities. Such an interpretation is not correct because of following reasons:

The concept of a spacetime singularity involves geodesic or worldline incompleteness, i.e., the worldline gets suddenly terminated at a finite value of the affine parameter characterizing it. This means that a parameter such as the comoving proper time or proper length of the curve would be finite even when a test particle would hit $R = 0$. But, we have found that there is no geodesic incompleteness for the formation of a zero mass BH. If one uses the O-S model, one would directly find that $\tau_c = \infty$ for $M = 0$. In a toy model involving collapse of an idealized Fermi-Dirac gas, it was explicitly found that[3], the proper radial distance $l = \int \sqrt{-g_{rr}} dr \rightarrow \infty$ as $R \rightarrow 0$. In general, since for material particles, $ds^2 > 0$ (except at central singularity), the singularity state of a BH (albeit with $M = 0$) cannot ever be reached; in other words, one must have $\tau_c = \infty$.

In contrast, a Naked Singularity is formed in a finite proper time and in case it is globally naked, there could be emission of energy even after its formation. (But since naked singularities are supposed to have $M = 0$ too, actually, there cannot any emission of energy from them in order that M remains non-negative.)

However if a globally naked singularity formation would be assumed, in the absence of a preceding Apparent Horizon and EH, the collapsing fluid would emit much more energy than would be available in a BH formation scenario. This is true for the formation of zero mass BH too and, in this sense, there is physical similarity between the two scenarios.

It has become almost a matter of ritual for many articles on BH and GC to refer to the so-called "Cosmic Censorship Conjecture" of Penrose[70] as one of the most outstanding "unsolved problem of classical GR" as if the much more fundamental question of whether trapped surfaces are formed in realistic collapses or not has been settled in favor of the former possibility. Note that the basis of this conjecture lies in the presumption that, in the first place, gravitational collapse of massive bodies would yield a singularity. Our work clearly shows that, this very presumption is incorrect, collapse of isolated bodies continue for infinite proper time without the formation of any singularity. However, mathematically, a spacetime singularity would form after infinite proper time and this singularity would be covered by an EH.

14 How Are BHs Surviving

The preprints by Mitra showing the non-existence and non-formation of BHs were first posted in Los-Alamos Preprint Archive way back in 1998[57,58]. A very long and quite raw preprint (though conclusions were correct) was also posted in the same arxiv[59]. An abridged version of this preprint got published in Foundation of Physics Letters in 2000[2]. In 2001, the same conclusion was rederived by Leiter & Robertson[32]. Further the prediction by Mitra that the so-called BH candidates are Eternally Collapsing Objects with strong intrinsic magnetic fields rather than any EH was tentatively found to be correct by a new analysis of observed properties of BHCs[67]. Subsequently much more work in this direction was carried out by Robertson & Leiter who explained many of the hitherto unexplained (such as a universal correlation between the radio and X-ray flux of both accreting neutron stars and BHCs) in a framework where the BHCs have physical surface and strong virialized intrinsic magnetic fields[71-75]. Such ECOs were termed as Magnetospheric ECOs or MECOs which are believed to be in quasistatic secular equilibrium by means of radiation and magnetic pressure and have extremely strong (but finite) surface gravity with $z \sim 10^{7-8}$.

However most of the other authors, institutions, conference organizers and following the average readers continued to ignore these works and the all pervading reign of BHs continued unabated. Why and How? Part of it is due to scientific reasons and part of it is due to non-scientific reasons. In order that future young readers can form an unbiased objective opinion on this fundamental problem of gravitational collapse, we would touch upon both these sectors:

14.1 Scientific Reasons for Believing in BHs

14.1.1

In Newtonian physics, the concept of a "Point Mass" or Massenpunkt with finite gravitational and baryonic mass (not just a test particle with $\delta M \rightarrow 0$) seems to be natural and happens to be the starting point of the formulation of mechanics and kinetic theory. This concept is so much ingrained in our psyche that it seemed most natural to carry it forward to GTR too. The fact that a finite mass uncharged Massenpunkt itself could be a physical singularity irrespective of whether it would give rise to BHs or not was not borne in mind by anybody including Einstein.

In hind sight, we can see now that *BHs are there because we assumed them (in the form of massenpunkts)*. However note that unlike the formulation of Newtonian physics and gravitation, the Einstein equations are field equations involving hydrodynamics of components continuous energy momentum tensor rather than discrete "Point Particles"!

14.1.2

Once one accepted the apparently most natural idea of a "Massenpunkt" (though GTR equations really did not require it because of its field theoretic/hydrodynamic nature), the beautiful Hilbert solution and attendant exact derivation of Kepler's 3rd law gave a strong support for the concept of BHs. The subtle point that Kepler's 3rd law involves finite mass finite radius non-singular spherical body (at least for field theoretic GTR) and not finite mass "point masses" was not noted by anybody.

Since Einstein was a physicist rather than an applied mathematician practising GR, his objection to BHs was primarily based on the physical notion of speed. And though Einstein's contention that BHs cannot be formed in order that material particles do not acquire the speed of light is very much correct, Einstein' objection to BHs was seen to be inconsistent because of the following implicit reasons:

The work of OV[25] gave rise to the idea that compact objects must have a finite upper mass limit and the work of OS[28] apparently indicated that GC could result in finite mass BHs. And Einstein did not directly address these topics. Furthermore Einstein did not raise any objection to the introduction of Point Mass Singularities which are practically synonymous with BHs by virtue of the mathematically correct Hilbert solution.

Nonetheless, to the knowledge of the present reviewer, none of the relativists believing in BHs and implying that Einstein's paper[30] was incorrect, ever showed it to be so.

14.1.3

Although all GTR workers, in principle, know that the externally perceived radius R_0 of the collapsing body is not the proper radial length, Newtonian intuition fails to recognize this and one unmindfully visualizes R_0 as the radius of the body (recall Joshi et al. referring to R as "Physical Radius"[61-63]). In other words, one, unmindfully, perceives that the radius of the body is necessarily finite. If so, it would appear that, if the collapse proceeds with some finite 3-speed, it would seem most natural that the collapse would terminate at $R = 0$ in a finite proper time.

On the other hand, we have found that during GC, the gravitational mass would go on decreasing and, yet, the field strength (both acceleration and spacetime curvature) would ever increase. This would keep on enlarging the proper radial depth (stretching of spacetime membrane) and collapse would continue indefinitely in the inner (proper) spacetime even though, externally, the mouth of the potential well $R \to 0$[2,3,4,5]. The equality of Eq.(198) can be satisfied only at this final stage, i.e., an apparent horizon can form as $R \to R_H \to 0$ and $M \to 0$.

Though, for a non-stationary metric, extended proper length may not be strictly defined, the actual proper radial length would be something like

$$l \sim \int_0^{R_0} \sqrt{-g_{rr}}\, dr \qquad (226)$$

The possibility that as $-g_{rr} \to \infty$ (near the would be singularity) l may diverge eluded everybody's intuition. But we have found that this is exactly what must happen because one must have $\tau_c = \infty$ (in order that the limit $2M/R = 1$ is never attained and correspondingly $ds^2 = 0$ is never achieved for a radial worldline for any finite proper time).

This picture is corroborated by a recent work which shows that for spherical gravitational collapse, it is possible to have a situation where[76]:

" the proper distance in the transformed metric from points 'near' the horizon to the horizon itself becomes infinite. However, the area of the spheres ($4\pi R^2$, $R \to R_H \to 0$, author) does not change (R hovering around 0) because the angular metric components are unaffected. This means that the manifold 'near' the horizon gets transformed into an infinitely long cylinder (inner spacetime, author) whose cross section asymptotes to the original area of the horizon and the 3-scalar curvature along the cylinder is a positive constant ($\sim 1/R_H^2$).."

As $R \to R_H \to 0$, the Kretschmann scalar blows up. And as the proper radial depth $l \to \infty$, the local observer's proper time of collapse $\tau_H \to \infty$ and the collapse becomes Eternal. Tomozawa has also shown that there is no EH at a finite proper distance[77] which too means that collapse must be eternal with the horizon forming only asymptotically after $\tau_H = \infty$.

However in a finite proper time, there would always be a finite M and finite R associated with an Eternally Collapsing Object (ECO).

Thus in GTR there cannot be any point singularity, a conclusion, which is in fine tune with the basic field theoretic (though classical) formulation of GR. It may be recalled that some of the quantum field theories like the string theories start with rejection of point singularities. But the fact that this exciting singularity free character (even at a non-quantum level) is naturally incorporated in GR has not been recognized by anybody. Far from it, for the last 50 years, GTR has suffered from a series of retrograde research glamorizing occurrence of singularities.

Hence, in contradiction to the mental picture of the GC and BH formation depicted in many books (MTW, p.855-856[17]) where the spacetime membrane suddenly gets pinched off, actually, the same gets stretched indefinitely and the collapse proceeds eternally in the indefinite gravitational pit.

This can be promptly cross checked by this simple argument:

If a GC would produce a finite mass BH, the proper radial length between the EH and the central singularity R must be finite in order that these two regions are separated by finite affine length. But on the other hand if an astrophysicist would free fall towards $R \to 0$, EH (MTW, p. 862 [17]), his length

$$l_{astroph} \propto R^{-1/2} \to \infty \tag{227}$$

But how can an astrophysicist *with length* $l_{astroph} \to \infty$ *be accommodated within a region having* $l = finite$. Would MTW elucidate? MTW would not and cannot elucidate this aspect because there is a fundamental inconsistency which has resulted from the wrong assumption of $\alpha_0 > 0$ for a Massenpunkt.

14.1.4

The chain of research by Anderson, Stoner, Chandrasekhar, Landau and Oppenheimer and Volkoff consolidated the notion of an "Upper Mass Limit" of baryonic compact objects. The idea of a compact object with a physical surface was that it must correspond to a stable and strictly static configuration. Once such an Upper Mass Limit got imprinted in the mind of both relativists and physicists, the question naturally arose that sufficiently massive objects would collapse beyond a state of static WDs or Neutron Stars or similarly massive other baryonic objects. And since for stellar mass objects, the upper mass limit suggested by standard nuclear physics (with central densities comparable to nuclear densities) lies well below the observed masses of many BHCs, it seemed most natural to think that collapse of sufficiently massive stars would end up as BHs and the observed BHCs are indeed such BHs (naked singularities have $M = 0$ and hence cannot represent the BHCs).

As is well known, a *static* spherical body has an upper limit of surface gravitational red-shift $z < 2[33, 34]$. Therefore it seemed even more natural that if the collapse would fail to stop at the neutron star stage with $z \sim 0.1-0.2$ and once the collapse would proceed to a stage with $z \geq 2$, the collapse would be unstoppable and a BH must form.

Here it was overlooked that the Upper Mass of baryonic objects actually refers to (i) **cold** objects, i.e., objects with negligible radiation pressure and (ii) strictly static objects. In other words, it never occurred to any body that the (i) observed BHCs could be extremely **hot** and supported by radiation pressure and that (ii) they need not be in strict hydrostatic equilibrium and represent stable configurations and hence their z need not be bounded by $z < 2$ rule.

Incidentally for **hot** self-gravitating objects, *there is no upper mass limit*. For instance the theoretically sound supermassive stars could have masses as high as $10^{10}M_\odot$ or even higher[44].

14.1.5

The almost (mathematically) exact OS dust collapse conveyed a general impression that at least in the idealized case there is a theoretical evidence of formation of trapped surfaces and finite mass BHs. In the absence of any possibility of exact solution of real life GC and, yet, in the eagerness of having the satisfaction of finding solution of this intractable problem, it was wished that when a real fluid would collapse, it can be approximated as an adiabatic collapse where M would remain more or less fixed irrespective of cosmetic appearance of time dependence in terms of $Q(t)$. But since R would change monotonically and must approach zero for sufficiently massive bodies, then, independent of the details of EOS, there must be trapped surfaces and subsequently BHs/Naked Singularities. In fact, *under the covert assumption of constancy of* M, apparent horizons/trapped surfaces and even EHs could form at arbitrarily low densities of the collapsing fluid.

The singularity theorems which are based on very reasonable assumptions (except the formation of trapped surfaces) seemed to be quite robust and ensure formation of BHs (naked singularities are considered to be of only academic interest by most of the readers).

The most fatal missing link in this chain of thoughts is the segregation of GR from physics by almost completely ignoring the magical H-K process by which GC develops its own antidote, fresh internal energy and emission of radiation. If there is no emission of radiation, there is no GC either. Thus in the absence of a radiative flux in the treatment of Goswami, Joshi and most of the other authors[61-65], $M(r,t)$ actually remains fixed. And what is this fixed value? In the absence of H-K process, $M(r)$ can be finite if either (i) there is no collapse at all, i.e., $\dot{R} = 0$. If yet, collapse would be assumed to occur ($\dot{R} \neq 0$) in a mathematical sense, for physical consistency, one must set $p = \rho = M = 0$. From this view point, actually, a non-radiative uniform dust collapse ($p = 0$, $\rho = uniform$) appears to be physically much more consistent and where one does not find any "Naked Singularity".

14.1.6

Many GTR experts like MTW[17], either explicitly or implicitly justify the OS model because formally the resultant metric has the same form as a closed Friedmann universe. But authors justifying the use of OS model because of this apparent mathematical similarity between the two problems completely omit the fact that *the boundary conditions for the two problems are completely different and though the Friedmann model corresponds to uniform density* **the pressure of the fluid is finite and EOS is arbitrary**. In the Friedmann Universe, neither is there a boundary where the density suddenly becomes zero nor is there an "initial radius" R_i where the universe would be assumed to be at rest unlike the finite isolated OS dust ball. In cosmology, the universe is in free fall despite having arbitrary pressure and EOS!

14.1.7 Misconception about Comoving Coordinates

Suppose one defines a flat spacetime using comoving coordinates:

$$ds^2 = dt^2 - dr^2 - R^2(d\theta^2 + \sin^2\theta d\phi^2) \tag{228}$$

Since the test particle (or fluid element) here is at a $r = fixed$ location, does it mean that one must have $dr \equiv 0$ in the foregoing metric? This certainly cannot be the true, because otherwise one would have written

$$ds^2 = dt^2 - R^2(d\theta^2 + \sin^2\theta d\phi^2) \tag{229}$$

at the very beginning. And then even for a photon or for a supposed Tachyon, one would have had time like worldline with $ds^2 > 0$. In fact if one would start with an incorrect metric of the form (229), one would not be able to use the comoving coordinates for any hydrodynamical or for any other problem. Just like a coordinate grid laced with fixed observers at (various) $R = fixed$ locations is not meant to exclusively measure the motion of a given $R = fixed$ observer (for which $dR = 0$) the metric (228) is not meant to exclusively monitor the motion of a given $r = fixed$ particle (for which $dr = 0$). On the other hand, the given observer at a certain $r = fixed$ location is free to measure the motion of anything else. For instance if the comoving observer is considered as an experimentalist in a floating boat in a flowing river, though he would find his own speed to be zero, he would measure finite values of speed for other floating boats, the trees on the river banks or boulders fixed on the river bed. Note that even in the comoving metric, the coordinate, R is very much present and in fact the angular part is solely dictated by the $R = constant$ grids. A boulder fixed on the river bed would lie at a

$$R = R_1 = fixed \tag{230}$$

and not at $r = fixed$ location. Thus, as the experimentalist would try to measure the *speed of the boulder* using the r, R grid, he would find $dR = 0$ or

$$\frac{\dot{R}}{R'} = -\frac{dr}{dt} \tag{231}$$

Thus the speed of the boulder, *in comoving coordinates* (not of the comoving observer himself) is

$$v_{boulder} = \frac{dr}{dt} = -\frac{\dot{R}}{R'} \tag{232}$$

And the speed of the experimentalist w.r.t. the boulder is

$$v = -v_{boulder} = \frac{\dot{R}}{R'} = \frac{\partial R/\partial t}{\partial R/\partial r} = -\frac{dr}{dt} \tag{233}$$

Note that this simple result has actually got nothing to the intricacies of GTR, on the other hand, it simply requires, intricate thinking about elementary Galilean physics. Though Mitra's proof for non-occurrence of trapped surfaces needed Eq.(231), it was however not necessary to interpret v as any kind of 3-speed. Nonetheless, for the sake of physical interpretation, he did mention that v was the 3-speed of the fluid. Since Mitra did not explain why it was so, several readers including J. Ehlers[78] concluded that since dr must be zero in a comoving coordinate (forgetting that it is so only for a given particle) and Mitra's proof must be faulty and has been obtained by confusing the partial derivatives in Eq.(230) as total derivatives.

14.1.8 Confusion about First Integral of Motion

A test particle in the Hilbert spacetime admits a first integral of motion, which is simply a statement of the conservation of rest mass energy[33,34]

$$g_{ik}\frac{dx^i}{d\lambda}\frac{dx^k}{d\lambda} = m^2 \tag{234}$$

where $\lambda = \tau/m$ is an affine parameter. Some readers feel that this equation prohibits an initial $ds^2 > 0$ from becoming $ds^2 = 0$. Note that, by dividing both sides of this equation by m^2 it can be rewritten as

$$g_{ik}\frac{dx^i}{ds}\frac{dx^k}{ds} = \frac{ds^2}{ds^2} = 1 \tag{235}$$

Thus the Eq.(216) is actually an identity where ds^2 can be both positive or zero (or even negative). Thus occurrence of $ds^2 \to 0$ as $R \to \alpha_0$ does not violate the existence of the first integral of motion. Hence ds^2 *can change its character even when the tangent vector retains its character* if the problem is seen from purely mathematical angle. But physics wise, such an occurrence would signify that one is no longer treading in a non-singular spacetime. Recall, in STR, $ds^2 > 0$ for material particles only as long as one observes the golden rule $v^2 < 1$. But if one would artificially enforce a situation, when $v^2 \to 1$, then one would have $ds^2 \to 0$.

14.1.9 Misconception about Gravitational Mass

Many physicists and astrophysicists (not relativists) confuse gravitational mass of the collapsing body with its baryonic mass which remains fixed during the collapse process. Thus they find the idea that M remaining more or less unchanged to be sound and conclude that BHs must be of finite mass.

Further by seeing illustrations/cartoons of BHs in many articles and books, many physics students presume that BHs must exist and have finite mass. Moreover repeated media publicity by observational astronomers that they have detected a certain BH at the center of a certain galaxy and the accompanying illustrations lend the impression that when BHs can be photographed, they must exist and be of finite mass. In reality, all such claims of having detected BHs actually refers to have detecting massive central condensations which are called BH Candidates. No one has ever detected or photographed an EH since, by definition an EH cannot be detected because of its infinite redshift. Observational astronomers however seem not to be too unhappy at the success of this regular false propaganda. They would rarely admit that they have actually detected a BHC and not a BH.

14.1.10 Claim by Numerical Computations

While a neutron star has a modest surface redshift of $z \sim 0.1 - 0.2$, an EH has $z = \infty$. If one would be handling a regime of $z = few$, one would already be in a regime of extremely strong gravity and in practice, it is impossible to carry numerical computations in this regime without making number of covert simplifications. Further in GTR all sources of energy densities, pressures and motions couple to each other in an infinitely non-linear way, and therefore, unless one is considering the motion of a test particle (no coupling), in real life, it is impossible to carry out "fully general relativistic" computation numerically although contrary claims are often made. In particular, even the best calculations would assume the radiation transport process to be not so strong and employ various covert simplifications. Some of the simplifications employed in the n'th paper would be known only when the (n+1)th paper would be written. However for a given n'th paper, there would usually be an air of certainty and confidence about the eventual result. Most codes would crash for high z regions and particularly if the layers of physics simplifications are removed, even when they would always be infinitely away from the $z = \infty$ surface. Nonetheless claims would be made of having found BH i.e., $z = \infty$ surface or working right up to the EH.

Many high energy astrophysicists study the physics of accretion disks around compact objects. When they actually work $\sim 10R_g$ or even $\sim 100R_g$ away from the center of symmetry and even when some of them would employ purely Newtonian formalism or at best use a free particle "effective potential" to simulate GTR effects, almost all of them claim that they are working on BH astrophysics. To convince the unsuspecting readers that they are indeed working on BH astrophysics, they may attach a sketch of an accretion disk with a cut out center having only a black dot (BH picture). Nonetheless, such papers do create an impression in the mind of an average reader that BHs are very much real and almost laboratory objects.

Further many astrophysicists numerically study the spherical accretion process around compact objects and supposed BHs. And in the past 10 years there have been repeated claims, out of such studies, that the EH has been detected. The full complexity of the accretion physics, its opacity, viscosity, radiation transport, angular momentum loss, magnetic properties and further the overall binary physics (variable accretion rates) cannot be exactly simulated in real life even if one ignores all GTR effects. However some of these authors do admit that the speed of the accreting fluid at the EH is none other than the speed of light[79]. But they do not realize that the occurrence of $v = c$ is an invariant phenomenon in the sense that if $v = c$ is true in one coordinate it must be so in all other coordinates too even though $v \neq v'$ in general. These authors also appear not to be aware that $v = c$ implies that the worldlines of the accreting fluid becomes null at the EH and this cannot happen in a non-singular region of spacetime. They are not aware that, therefore, the very occurrence of $v = c$ at the EH means that the central object is not a BH of finite mass. If it is still to be fancied as a BH, its mass $M_0 = \alpha_0/2 = 0$! Otherwise the speed of the converging accretion flow would be $v > c$ once it would cross the EH.

Ironically it is such claims by numerical physicists which bolster the idea of the BH of relativists who would, however, like to claim that nothing special happens at the EH and speed of an infalling particle must be subluminal there!

14.1.11

Once the above mentioned misconceptions clad in the robe of apparently natural ideas got institutionalized after the death of Einstein, by and large GTR research got dominated by a new breed of charismatic researchers who were first mathematicians/applied mathematicians, then relativists and finally physicists. In contrast, Einstein was first a physicist, then a relativist and lastly a mathematician. Or more aptly, Einstein was first a physicist, then a physicist and finally also a physicist. An applied mathematician is eager to look for solutions of a given set of equations without having sleepless nights worrying about the validity of the assumptions needed to fulfill such a task. He is also eager and happy to apply newer mathematical tools like charts, diagrams, complex analysis, complex topological concepts in the context of the physical problem rather than repeatedly worrying about the necessity of such approaches. In particular, an applied mathematician need not worry too much about the following questions:

(i) why should R, originally defined as a spacelike variable, suddenly change its character when both the EH and the central singularity are still marked by $R = \alpha_0 = fixed$ and $R = 0 = fixed$ markers respectively rather by $T = fixed$ markers ? Note, even for the EH or the region interior to it, the invariant surface area of the 2-spheres continues to be $4\pi R^2$ and not $4\pi T^2$!

(ii) If a variable is originally defined as a real one (as all observables must be), a true physicist or a true pure mathematician will get anxious if the variable turns out be an imaginary quantity in a certain region. The most logical physical action, in such a case, would be to realize that one must be treading into a region which cannot form in real life. But the occurrence of a complex variable could be a matter of delight and a new impetus for imparting techniques of complex variables and even adding newer topological dimension into the problem. If a physicist or a true pure mathematician would be compelled to use a transformation $t \to it'$ in a certain situation (for purely mathematical expediency) he would be extremely careful in interpreting the results and ferreting out the physically meaningful *real* part out the complex mess. However, some of the applied mathematical relativists might be happy finding some new spacetime property out of those complex equations. In fact, in one of the approaches, the occurrence of Hawking Temperature is related to temporal periodicity in *imaginary time* obtained after $t \to it'$ transformation[16]!

(iii) A physical situation such as the static spacetime of a massenpunkt must always show the $x^0 \to -x^0$ symmetry and if a certain coordinate transformation would violate this basic symmetry then an applied mathematician need not worry too much. On the other hand, this violation would offer an opportunity for constructing new topology.

However if some of the physicists following the trail of this mathematical developments find it uncomfortable to have an imaginary quantity as a physical variable, he or she may just put a modular sign at an appropriate place without any mathematical justification. For example, when there is no modular sign in the worked out value of T in Eq.(25-38) of MTW[17], they added a modular sign in Eq.(31.2) (p. 820) and Eqs.(31.9) and (31.10c) (p. 824). Similarly, while there was no modular sign in the correct Eq.(25.3) (p. 663) of MTW, they quietly added a modular sign in Box (31.2) (p. 828).

Eventually such applied mathematician cum relativists would define the course of GR as a subject of physics. Subsequent physicists working on the same problem would find this modular signs to be convenient or meekly overlook this non-mathematical and non-physical action instead of being left aside. Fresh GR students would be fed with such a version of GR from the very beginning, and, soon, they would start defending such approaches which they have learnt.

(iv) Eventually a paradigm would be built which would be supported by a labyrinth of charts and diagrams (rather than by results obtained by exact solutions) and subtly inaccurate but convenient mathematical exercises which are essentially mental pictures obtained by implicitly generalizing idealized physical phenomenon such as OS dust collapse. In the absence of exact solutions, even the general physical situations would be explained in the backdrop of such charts and diagrams. Slowly over the years the distinction between real intractable physical problems and wishful mental pictures depicted by charts and diagrams would get blurred and often discussion would start from the charts rather than from the Einstein equations.

By and large, GR would then be pursued as a strange subject on its own right not necessarily organically connected to every man's physics. And if any reader/physicist would insist on going back to the original Einstein equations or associated measurable basic notions like velocity, acceleration, finiteness of scalars (at non-singular regions), he or she may be depicted as ignorant of the nuances of topology, differential geometry, branch cut and would finally be silenced by the lethal veto power of a supposed "coordinate singularity".

One example of abuse of topology for finding the spacetime of a massenpunkt is the Kruskal metric where in the excitement of having invented such a "maximal extension" it was forgotten that *the Kruskal metric is not even asymptotically flat*; $g_{UU} = -g_{VV} \neq 1$ at $R = \infty$ (they are zero). Yet all graduate students are indoctrinated into believing that the invention of the Kruskal chart is one of the pinnacles of GR research.

The purpose of this subsection was not to undermine the importance of mathematics in GR or any in any other branch of science. GR, because of its mathematical complexity required the services of mathematicians perhaps more than any other subject. Without their active participation GR would not have been developed at all, and the most important contribution to this effect was due to Grossman. Even now research in pure mathematics particularly in Differential Geometry is strengthening and enriching GR and QG. One of the best examples here would be the proof of positivity of mass[80]. Such pure mathematicians do not pretend to be physicists at all, some of them may not claim themselves even as relativists. They do not ever use any wishful EOS and solve Einstein equations to convey results which could be completely physically incorrect. They do not build an edifice built on such misleading solutions, they would never quietly put a modular sign ever to suppress the imaginary nature of a quantity.

14.2 Non-scientific Reasons

For last few decades, amongst all the research topics in physical science which have evinced overwhelming popular and media interest, probably, BH tops the list. Although it is often mentioned that theoretically BHs must exist and observationally they were discovered at least 20 years back (as the compact object in Cygnus X-1), in practice, almost every month there would be some news item announcing "discovery" of BHs. The aim of detection and study of BHs is one of the prime ingredients for asking for big grants for big experiments in optical, X-ray, $\gamma - ray$ and Gravitational wave astronomy. There would be hundreds of text books, thousands of articles and probably tens of thousands of conference papers discussing various aspects of BHs. Given this atmosphere, any author claiming non-occurrence of BHs would be considered a spoilsport or a sheer nuisance. UK's best known scientist Stephen Hawking became so because he is supposed to have shown that exotic BHs would be even more exotic by emitting radiation if one would apply Quantum Mechanic (QM) to them[81]. Then he became even more famous when quantum information physicist John Preskill (probably) first pointed out that Hawking radiation would entail loss of quantum information from the universe and hence is in contradiction with QM. But Hawking would not budge for 30 years and the problem rather than being solved after proper scientific methods remained alive to regularly kick the media headlines.

What can be seen from this discussion is that BH research has become so glamorized and a topic involving so much personal stakes of so many powerful and influential academicians and institutions that now it is hardly a topic of impersonal professional and analytical scientific debate. As if nobody wants to be left behind in having a piece of the BH pie, whether it is classical relativists, observational astronomers, numerical physicists, theoretical particle physicists, experimental particle physicists (recall the sensation following the news that particle colliders may generate micro BHs and which might slowly engulf everything around them). In this surcharged atmosphere, if anybody would show that there cannot be BHs, most of the referees would reject the manuscript with one liners like "there must be obvious mistakes in this claims and it is the author's responsibility to find out such mistakes.. and a referee cannot waste his/her time detecting such trivial errors.." or "I am not convinced by the author's rejoinder, and I see no reason to change my initial comments..". Some of the referees could be more generous and attempt to justify their rejections.

Some of the specific reasons offered to claim existence of BHs are:

14.2.1 Confusing Spacetime Filled with Mass Energy as Vacuum Spacetime

T. Padmanabhan is an excellent scholar (Inter University Center for Astronomy and Astrophysics, Pune, India) and author of a series of excellent books on astrophysics and cosmology. But like almost everybody else, he too is absolutely sure that BHs are real and GC would produce them provided one would use "proper coordinates":

As per him, one must treat the GC problem by using only the Kruskal coordinates[82]. But note that the Kruskal coordinates are meant to study the geodesic of a free falling test particle *in vacuum* and hence even for a free falling dust collapse one cannot invoke the Kruskal coordinates (except for the boundary layer of the dust ball). Obviously he is impressed by the hype about both the Kruskal coordinates and GC and cannot help connecting the two. And he declared that Mitra found non-occurrence of trapped surfaces in GC because Mitra used comoving coordinates rather than Kruskal coordinates (all workers in GC know that comoving coordinates are most appropriate for GC and OS too used them). The reason that such an excellent scholar is making such a comment is that first he assumes that trapped surfaces must form and one must invent some "error" in any work which does not fall in line. Second, if such a profound result must be correct, it has to emanate from abroad, preferably from non-Indian authors of the sufficiently high stature. On the other hand, Indian physicists working in India are by and large supposed to expound on new directions developed abroad.

D. Lohiya of Delhi university, a research worker in Quantum Gravity, and also a former graduate student of none other than Stephen Hawking (and also a friend of T. Padmanabhan), too went on record in a live TV interview/debate[83] that Mitra's proof is completely wrong because Mitra did not use (vacuum) Kruskal coordinates for studying GC of an arbitrary fluid! He further described Mitra as a "laughing stock" having "no idea of how to use coordinates in GR" despite Padmanabhan having taught him that GC must be done only with Kruskal coordinates!!

Fortunately Stephen Hawking's student did not suggest persecution of the non-believer.

14.2.2 Desperate Belief in Non-existent Proof of Trapped Surfaces

Abhay Ashtekar is a highly respected research worker and an authority in Quantum Gravity. He is also supposed to be an expert on GC. But he rejected[79] Mitra's proof on this ground: he declared that Eq.(5.2) of a certain paper entitled "Binding Energy for Spherical Stars" by P. Bizon et al.[85] has already proved that trapped surfaces do occur. But what was this equation? As any body can look up, the said equation is

$$M \leq 2L \tag{236}$$

In the first place L is not the invariant circumference coordinate (R) at all; on the other hand L is the "Proper Radius" of the star. So this was Ashtekar's first piece of lapse while evaluating this work. Even if one would fancy L as R, the Eq.(217), would lead to $M/2R \leq 1$, which would just look like the equation Mitra obtained for Non-occurrence of trapped surfaces. Thus Ashtekar, in his eagerness to reject Mitra's proof, nonchalantly, fancied the $\leq$ sign as ≥ 1 sign! Actually this entire paper by Bizone has got nothing to do with GC because it deals with static self-gravitating bodies and no collapsing bodies at all. This is a classic example of what prejudice can make out of even bright scholars. And interestingly, the Editor too argued that Ashtekar was correct, i.e., on certain occasions, $\leq$ sign is to be read as $\geq$ sign; a non-collapsing problem static problem could be considered as GC problem!

14.2.3 Confusion about Notions of Speed in GR

As we have discussed in Section (10), in the absence of a spacetime cross term in the metric, the general expression for 3-speed gets simplified. Such a simplification is possible when (a) the non-stationary metric is diagonal and trivially (b) if the metric is static. While discussing a static spherical metric (e.g., a test particle around a spherical body), Landau & Lifshitz accordingly used the Eq.[8,9]

$$v^2 = \frac{-g_{rr} dr^2}{g_{00} dt^2} \tag{237}$$

And when one is dealing the internal motions of a spherical fluid, the same expression would hold good and in such a case (because of the absence of spacetime cross term) even if there would not be any static metric at all. Note here that our proof of non-occurrence of trapped surfaces does not even require the definition of 3-speed, and 3 speed is mentioned only for physical interpretation. As discussed earlier, v here is the speed of the fluid w.r.t. a $R = constant$ marker.

In contrast to the above attitude, the Board of Editors of *Annals of Physics* did a much better job when it decided to reject the manuscript[59] with the following candid email [86]:

```
From: annals@MIT.EDU Wed Mar 24 02:43:22 1999

Subject: Re: Appeal on AOP64317

Dear Dr. Mitra,

On behalf of Dr. Feshbach and the editorial board,
I am writing to let you know that they are not able
```

```
to reconsider this paper.  They are not able to publish
every correct paper and have made their final decision.
Please do let me know if you would like the manuscript
forward to another journal or returned to you.

Yours sincerely

Eve Sullivan

for the Annals of Physics
```

14.2.4 Sheer Academic Intolerance

Since Mitra put up his preprint showing non-occurrence of trapped surfaces and BHs in 1998, hardly any rebuttal of it has so far appeared either as a preprint in LANL or in any refereed journal. However even if it were not possible to do so, John Baez (University of California, Riverside)[87] and Chris Hillman[88] (having no affiliation with any academic institution) launched a vicious personal attack on Mitra by primarily (ab)using an internet newsgroup

http:// www.sci.physcis.research

run by themselves:

John Baez's personal webpage contains links to various notes on GTR including one called "wrong.html"[89] which is actually authored by Hillman and was originally featured in a section titled "Hall of Shame". Without going into the details of this murky postings by John Baez who as an academician should have published his scientific critique, if any, in appropriate refereed journals, it is only pointed out that, neither he nor his friend Hillman actually pointed out any error in the proof showing non-occurrence of trapped surfaces. On the other hand, they kept on harping on the known standard works/ideas on BH paradigm built by **assuming the integration constant** $\alpha_0 > 0$ rather than by fixing it by physical requirements (divergence of Acceleration Invariant at $R = \alpha_0$). It is only recently that in response to discussions in another Newsgroup

http:// www.sci.physics.relativity

John Baez has attempted[90] to put up his precise objection to Mitra's work:

" Starting from the solution which describes a black hole of mass m, he attempts by a calculation to show that $m = 0$. It's bit like taking an arbitrary prime number and proving that it must equal 37."

Note that when an integration constant is obtained in a physical problem it is really not arbitrary and one must not a priori assign any sign or any numerical value to it. One must obtain the actual value (both sign and number) by using either boundary conditions or given initial conditions involving physical quantities. Thus, no integration constant in physics is really arbitrary even though it might symbolically look so. The physical condition we employed in fixing α_0 is that all local scalars involving the Riemann tensor or the metric coefficients can diverge only at a genuine physical singularity, i.e., at $R = 0$. John Baez and many other with their completely prejudiced mind simply would not try to appreciate this mathematical and physical reasoning.

In Section(11) we found that the occurrence of a physical singularity at $R = \alpha_0$ is also manifest by occurrence of $ds^2 = 0$ for a radial geodesic. And since $ds^2 = invariant$ this result must hold in any coordinate system including the most innocuous free falling Lemaitre frame. This direct proof has made John Baez uncomfortable and fortunately again he has invented a "reason" behind this result[91]:

"in Lemaitre coordinate r is a constant for such a particle, so, $dr = 0$ and hence $dr = dt$ cannot be true."

We mentioned in detail that there is lot of misconception about the comoving coordinates even amongst many noted scholars. In particular, we obtained $dr = dt$ by comparing Eqs.(158-159) rather than by doing any division by 0. But Baez's confusion about comoving coordinates is too evident here.

As discussed, in the comoving metric $dr \neq 0$ in general (otherwise the metric itself would be meaningless without a spatial part). And even if dr would be zero, then too, Eq.(161) would be correct, because $\lim x/x = 1$ even when $x \rightarrow 0$ or anything else.

The above discussion clearly shows that scientific reasons (misconceptions) apart, there is an organized attempt, by using various non-scientific means, including personal vilification, to suppress scientific truths which do not pay obeisance to this BH paradigm.

John Baez's webpage also has a file called "history.html"[92], which is again authored by his close professional associate Chris Hillman. Fortunately this html file happens to be, in general, a meticulous account of the development of the false BH paradigm. And though the BH believers routinely and unethically use Einstein's legacy to further strengthen this false paradigm, it is well known that though Einstein could not self-consistently show that BHs are unreal, nonetheless, he never accepted this paradigm. And therefore Einstein too could not escape the contempt of John Baez and Chris Hillman:

In 1939, Einstein publishes a paper which presents a rather desperate (and entirely incorrect) argument that no body could collapse past its Schwarzschild radius. The nature of the conceptual errors in this paper show that Einstein still did not understand either the distinction between a coordinate singularity (the boundary of a coordinate chart)and a geometric singularity, nor the distinction between local and global structure. (Indeed, there is no evidence that Einstein ever understood correctly the geometry of all exact solutions to his field equations.)

This quote unambiguously shows that the so-called GTR experts have made a make believe BH paradigm by dissociating mathematics and geometry from physics (feigning finite density when pressure is zero, overlooking divergence of physical scalar a at the EH and so on) and would not spare anybody pointing fingers at this glass house.

15 Quantum Mechanics and BHs

It is often claimed that the most promising route to Quantum Gravity (QG) is through the Super Symmetric String Theories. Probably it might be so. And if a given string model is correct, it should (i) either yield no neutral BH or, if at all, (ii) yield only $R_g = 0$, i.e., zero horizon area BHs. In fact many of the string models indeed yielded extremal BHs[93-95]. To see some of the original results of String Theory *vis-a-vis* BHs, let us heavily rely on a 1992 review by Horowitz[93]:

For a charged extremal BHs, the proper distance to the horizon $R = M$ from a point $R = R_0 > M$ along a $T = constant$ radial curve is

$$L = \int_M^{R_0} \frac{dR}{1 - M/R} = \infty \tag{238}$$

Thus many string theory BHs indeed yield the infinitely long pit as we found using classical GR (for neutral BHs). And these extremal BHs indeed have zero horizon area and zero Hawking temperature, i.e., there is no Hawking radiation at all. In the case of uncharged BHs, one would have $M = 0$ and in which case one would obtain exactly the same picture found for the GR case described in this review.

Also the solution for the so-called *Straight Strings* have **no** EH. Simultaneously, in the extremal limit, the *Black Strings* become *Straight Fundamental Strings*. Further, as summarized by Horowitz,

"For black holes, one of the most important (property) is that the string metric describing an extremal magnetically charged black hole has neither a horizon nor a curvature singularity. The spacelike surfaces contain infinite throat."

And some of the "open problems" for the string theory as mentioned by Horowitz in 1992 are the following:

- "It would be of great interest to find the exact solution to string theory which approaches Schwarzschild when the higher order corrections to the field equations become unimportant."

"Although new exact solutions have been obtained this way, so far *none can be interpreted as an asymptotically flat four dimensional black hole*"

- "If one applies the transformation (2.13) to time translations in Schwarzschild, one obtains another solution in which *the horizon becomes a singularity*"

- "Finally, there is the problem of calculating the Hawking evaporation of black holes and black strings in string theory. This appears to be beyond our current ability, although progress is being made."

Similarly, in 1995 too, String Theorists were getting zero mass BHs aplenty, but were confused as to what to do with them. Part of this confusion arose from their attempt to model the elementary particles as BHs! This is really ironical because when string theory is tailored to banish singularities, in practice, many string theorists are attempting to model the elementary particles as singularities. As Strominger writes[94]

• "The consistency of our picture requires that there is one and only one supermultiplet which becomes *massless at the singularity*."

"At first it may seem surprising that classical black holes can be *massless*. However this phenomenon has an appealing explanation from a ten-dimensional perspective. The IIA (IIB) theory has *extremal* black twobrane (threebrane) solutions whose mass is proportional their area. After Calabi- Yau compactification these may wrap around minimal two (three) surfaces and appear as four-dimensional black holes. As the area of the surfaces around which they wrap goes to zero, the corresponding *black holes become massless*."

All the emphasises in the above quotes are due to this reviewer.

What transpires from these quotes is that String Theories may be strongly signalling that for overall consistency only zero mass or extremal BHs are to be allowed. However, QG researchers have all along been misled by the false classical BH paradigm and they have been very unhappy with this result that $S_{BH} \equiv 0$. Accordingly, they have been struggling to generate countable "hairs" (i.e., QM states) to the BHs to chase the chimera of "finite mass BHs". Also, from a very fundamental consideration, QG should not yield any nonextremal BH because of the following reason:

Since the macroscopic BHs are necessarily classical objects and have no "hairs"[17], their quantum state is uniquely determined by only Mass, Charge (if any) and Angular Momentum (if any). This implies that the entropy of a BH is unique $S = \ln 1 = 0$. And since BH entropy is related to the horizon area A as[96]

$$S = \frac{A}{4G} \tag{239}$$

QM demands a unique value of $A = 0$, i.e., $R_g = \alpha_0 = 2M_0 = 0$, for neutral BHs or probably $M = Q$ for charged BHs. Thus we see that, actually, the classical GTR result of $\alpha_0 = 0$ is in perfect agreement with the QM picture.

As Mathur[96] discusses, the **D1-D5 solutions have no horizons and no singularity**. Recall that though the aim of all physical theories is to avoid singularities the String Theorists are so much charmed, overawed and misled by the classical BH paradigm, that they are worried at such non-occurrence of horizons and singularities. They think, if their theory in on correct track, first, they should obtain horizons, singularities and finite mass BHs! Then they must obtain the "hairs" which would give rise to supposed (actually fictitious) BH entropy. Further, they also must obtain Hawking radiation, then demand that Hawking Radiation would be in conflict with QM and Quantum Information Loss Paradox would be highlighted/created. Only after having done all these, they would try to resolve the conflict between QM and GR!

In other words, even when, at least on physical grounds, QG may not have any finite area/entropy BHs, they must invent one to ensure that their theories yield proper GR behavior in the low energy range. Clearly QG research has been perfectly misled by the false BH paradigm.

Without pretending to give either any authoritative or comprehensive review on BHs in String Theories (this author actually does not know any Quantum Gravity) we would briefly do some piece meal job. In 1995, Sen proposed to have non-extremal BHs with finite entropy by introducing the so-called *Stretched Horizon*[95]:

• "To resolve this puzzle, we could postulate that the entropy of the extremal black hole is not exactly equal to the area of the event horizon, but the area of a surface close to the event horizon, which we shall call the 'stretched horizon'. This assumption is not totally unreasonable, since for various reasons we expect that our standard understanding of the physics of the black hole has to be modified very close to the *horizon/singularity*; and the stretched horizon represents a surface beyond which our standard understanding breaks down."

This whole scheme is completely *ad hoc*, inconsistent, and **incorrect**. Unless and until we recognize the importance of acceleration as an independent measure of the gravitational field, the gravitational field at the EH can be made arbitrarily small (unless $M_0 \equiv 0$) because as per standard GR interpretation, it is only the Kretschmann scalar ($\sim M_0^{-4}$) (see Eq.[4.3]) which is the measure of gravitation (if z would really be accepted as the measure of strength of gravity, then, by definition, the EH with $z = \infty$ would be admitted as the real physical singularity). And in any case, if BHs are admitted, there is no question of "our standard understanding" breaking down because the *EH is believed to be an absolutely regular and non-special region of spacetime. The occurrence of a "coordinate singularity" at the EH is considered to be as commonplace as the occurrence of a singularity in $g_{\phi\phi}$ at* $\theta = 0$. But, note that Sen virtually equates the EH with the central singularity by using "horizon/singularity" phrase.

(In fact the present author had sent two detail emails to Sen explaining why the horizon and the singularity are indeed synonymous and why there cannot be any finite area horizon; but Sen never acknowledged those mails).

So Sen's proposal is in absolute violation of GR unless $\alpha_0 = 0$ but which he would not admit. And if GR can be really violated, then why must we worry about BH entropy and Hawking radiation?

To define the location of this adhoc "stretched horizon" Sen adds[90]

• "If we *believe* that the necessary modification of the physics near the horizon comes from the string world sheet effects, then we should define the stretched horizon to be the surface where the string world sheet theory *becomes strongly coupled*. In other words, this is the surface where the *space-time curvature* associated with the string and/or other target space field strengths, becomes *large*."

This whole proposal shows lack of understanding/disregard about classical GR. The Kretschmann scalar, the measure of the "space-time curvature", at the EH, $K = (3/4)M_0^{-4}$, *can be made arbitrarily small* and further any local acceleration can be made to vanish after suitable coordinate transformations (unless one realizes that $M_0 \equiv 0$). Thus let alone any "stretched horizon", the "horizon" itself and even all the region interior to the horizon (except the central singularity) must be completely regular spacetime without any physical abnormality (if $\alpha_0 > 0$). And there is no question of a preferred region (stretched horizon with special properties) at or above the horizon.

Thus eventually, like the GR itself, string theory too may have suffered from regressive research by chasing the chimera of finite area BHs.

Therefore, unless one recognizes the EH itself as the central singularity (as is indeed suggested by many models of string theory) invention of a "stretched horizon" with special properties is an unacceptable doctoring of physics.

Afterwards what follows is series of approximations and adhocisms to obtain the classical (finite area) BH entropy formula. So though BHs have "no hair", hairs must be invented to pursue a red-herring even though the eventual result, correct or incorrect, would be self-contradictory - having both "no hair" and "sufficiently many hairs! If a person is bald, he is so independent of the theories the observers may employ. At the most, an observer with a microscope may find some microscopic strands. But no one can claim to find even a single macroscopic strand of hair on a completely bald head.

Hawking, and following him, hundreds of other workers claim that if QM would be applied to the problem of GC, the resultant BH would radiate[81]. This cannot be true in general if any Quantum Gravity effect would ever be responsible for such particle production because, on the EH, $K \sim M^{-4}$ and can be arbitrarily smass for sufficiently high M. In any case, such particle production is not Hawking radiation. On the other hand, Hawking radiation is a net particle flux seen by a distant observer at late times (due to infinite time dilation at the EH) specifically because of formation of an EH. Recall that, even if there would be no particle production due to any quantum gravity effect, good old H-K mechanism requires that, there is no GC without accompanying emission of radiation. Thus whether one considers quantum effects or not, for a truthful description of the GC process, in principle one must track the ever decreasing value of M. Did Hawking or any of his innumerable followers do this before claiming the existence of Hawking radiation? The answer is a "no"! Let alone attempting to track the decreasing value of M, most of them do not even recognize its necessity. Implicitly, all of them including Hawking, assumed M to remain constant despite the explicit consideration of a radiative flux. On the other hand, they considered that M would decrease only after a (finite mass) BH is formed.

Many of the Quantum Gravity (QG) workers may not even be aware of the H-K mechanism. They may not be aware that when they talk of Hawking radiation, they are also treading onto the (preceding) classical GC phase and Hawking radiation, if at all, has to be seen in the perspective of GC. However, in QG context, one talks of "back reaction" without realizing that QG effect apart, classical gravity demands "back reaction" in the form of H-K process. Further, it is also presumed that the effects of the "back reaction" would be minor and non-qualitative. However Hans Stephani, in the concluding section of his excellent text book[33] did show a rare foresight and momentarily discussed the possibility that **the eventual mass of the BH formed in GC could be zero due to "back reaction"**.

Our work has shown that, irrespective of the existence of speculative, QG back reaction, there must be "back reaction" of purely classical and thermodynamic origin during GC: and it is this classical "back reaction" (emission of neutrinos/photons), the eventual mass of the

BH must be zero. Since only uniform dust collapse can be analytically handled there is no question of truthfully analytically tracking the evolution of M during GC. In that case, is there any necessity, to predict novel results associated with GC? Without insisting for an answer of this question, let us see whether, in the first place, QM or QG, actually, allows the existence of BHs or not.

15.1 Quantum Upper Limit on Proper Acceleration

Quantum mechanics sets a fundamental unit of proper length called Planck Length

$$l_p = \sqrt{\frac{\hbar G}{c^3}} \tag{240}$$

and a fundamental unit of proper time called Planck Time:

$$\tau_p = \frac{l_p}{c} \tag{241}$$

Accordingly, QM sets an upper limit of *Proper Acceleration*

$$a_p = \frac{Maximum\ Speed}{Minimum\ Proper\ Time} \sim \frac{c^2}{l_P} = \frac{c^{7/2}}{\sqrt{\hbar G}} \tag{242}$$

Detail considerations yield[97]

$$a_p = \frac{c^2}{2l_P} = \frac{c^{7/2}}{2\sqrt{\hbar G}} \tag{243}$$

Thus *QM does not allow occurrence of infinite proper acceleration*. But we have found that, as long as the external spacetime of a collapsed body is considered as vacuum, the Invariant Acceleration (and also the proper acceleration) of a test particle lying on the surface of the collapsed body is

$$a = a^R = \frac{GM}{R_0^2\sqrt{1 - 2GM/R_0c^2}} \tag{244}$$

Thus QM would not allow any object to collapse to a radius smaller than R_{QM} determined by the following Equation:

$$\frac{GM}{R_{QM}^2\sqrt{1 - 2GM/R_{QM}c^2}} = \frac{c^{7/2}}{2\sqrt{\hbar G}} = Finite \tag{245}$$

While classical GTR demands that $M \to M_0 = 0$ as $R_0 \to \alpha_0 = 0$, QM ensures that $R_{QM} = R_0 > \alpha = 2GM/c^2$ (even if one would consider $M \to 0$ limit) and no EH is ever formed. Hence the question of attaining the central singularity at $R = 0$ simply does not arise in QM. If QM is really works at a fundamental level, it must allow macroscopic classical ECOs if the effect of QM/QG may not be visible. Numerical solution of Eq.(245)

indeed shows that it allows arbitrary high finite value of M for collapsed objects having radius extremely close to but higher than R_g. Thus QG/QM indeed allows existence of ECOs. On the other hand, the minimum value of $R_{QM} \sim l_p$. This suggests that all the QG tells that there are ECOs but their minimum size is not zero but l_p.

And as we discussed, some of the string models indeed have solutions having no horizons and, in general, finite area, BH solutions may be altogether absent. But it is really unfortunate that String Theorists, instead of picking up these correct messages, are struggling to invent finite mass BHs and its associated baggage of physical inconsistency.

In the preceding sections, the idea of ECOs was obtained without QM. It was seen that in case if one would imagine the collapse to proceed up to $R = 0$, it would require an infinite comoving proper time to reach the massenpunkt state at $R_0 = 0$. Note that it was never insisted that such a massenpunkt state with $R_0 = 0$ will be actually reached. And even if a massenpunkt will appear to form externally, the proper radius of the massenpunkt would be $l \rightarrow \infty$, i.e., there would be an infinite "inner spacetime"(infinitely long "throat") and no point singularity.

We see that QM leads to a more definitive picture where GC cannot ever reach the massenpunkt state at $R_0 = 0$, on the other hand, one must have, $R_0 > R_{QM} > l_p$

So we clearly see that if QM is to be applied to the gravitational collapse problem, in order that proper acceleration of the collapsing fluid remains finite, QM, in the first place, would not allow formation of any BH nor any naked singularity. During the formation of the this QECO, there can of course the emission of radiation by both classical H-K mechanism and Quantum Gravity effects but there would not be any late "Hawking Radiation" because there would not be any EH at all. Hence we find that the entire idea and calculations for supposed "Hawking Radiation" was yet another example of building castles in vacuum by imagining formation of finite mass BHs in violation of both GTR and QM.

Quantum Gravity workers, in general, may not be aware that the Acceleration Invariant would diverge at $R = 2M$ in violation of QM. On the other hand, they naively believe that the existing classical BH paradigm is correct where, necessarily, BH mass $M_0 > 0$. However, in the QG model proposed by E.R. Caialiello[98], an upper limit to proper acceleration does exist: When the acceleration induced by the background gravitational field is large enough, the elementary strings would be subjected to Jeans like instabilities and string extremities would get causally disconnected.

16 Quantum Information Paradox?

If a finite mass BH would exist (in violation of QM and GTR), as per Hawking, it would keep on radiating and eventually its mass $M \rightarrow 0$ and radius $R_0 \rightarrow 0$. Since this would happen at a finite value of proper time and affine parameter, (the possibility of proper radial distance $l \rightarrow \infty$ would be absent in this case when there is already an EH), it is believed that the BH would "vanish" (where ?) from our universe. Accordingly, all the information content originally trapped within the EH too would vanish from the universe in violation of GTR and QM. John Preskill was probably the first person to raise legitimate objection against this scenario. Obviously, the stumbling block in non-preservation of Quantum Information is the supposed existence of the EH. However since QG physicists too have accepted the BH paradigm there cannot be any convincing and transparent resolution of this problem in the existing framework.

Some of the QG treatments not resorting to inconsistent ad-hocism of stretching the EH, may indeed find that GC, instead of BHs, would rather, generate, UCOs with radius $R_0 > R_{QM} > R_g$. It is likely that the "fuzzball" suggested by Mathur[96] could be such an object. Unfortunately, it is not so. And, a statement such as "BHs are actually fuzzballs" is completely incorrect. Thus Mathur's view of

(i) First accepting the finite area BH paradigm where all the mass energy is concentrated at the central singularity

(ii) and then imagining a picture where the empty region between the EH and the central singularity is filled with a tangle of strings (to generate the quantum states necessary for explaining fictitious BH entropy)

(iii) and then working out the Hawking radiation

is a tangle of contradictions because a BH is a BH with an EH which is the boundary of a trapped region with infinite radial acceleration and infinite z. In such a case, nothing, neither strings nor outgoing photons can stay put at or within the EH unless $R_g \sim l_p$. In other words, a BH can be seen as a "fuzzball" only by "fuzzying" the GR.

And if a fuzzball is to be real, i.e., it has to be a non-BH though its radius could be extremely close to R_g. In such a case, there would be no question of any Quantum Information Paradox in the first place and there would be no need for any solution to the fictitious paradox either.

Incidentally, in 2004, Hawking retracted his 30 year old view and has opined that BHs do radiate everything including energy and information[99]:

"Information is lost in topologically nontrivial metrics, like the eternal black holes. On the other hand, information is preserved in topologically trivial metrics"

So now as per him the spacetime around a "point mass", considered by Hilbert, Schwarzschild and many others 90 years back, could have completely different "topologies"! Why? And which one is the real BH? What is precisely meant by "trivial" topology and why should the BHs he considered for 30 years should have "non trivial" topology? And how does the transition from "non trivial" to "trivial" topology take place? Are there only these two extreme vagaries of BH topology or there is a full spectrum of gray areas with various sheds of gray BHs which simultaneously radiate and gulp Quantum Information. Since Hawking has not made a preprint of this research available (Oct. 10, 2005), probably nobody knows precise answers to the questions posed here. Were there not sufficient discussion or debate over his claim[99]

> "I think I have solved a major problem in theoretical physics since I discovered that black holes radiate thermally, thirty years ago."

What was the precise drawback in that 30 year old discovery? It turned out that there was no meaningful discussion because as John Baez writes[100]

"Stephen Hawking decided to announce his solution to the black hole information loss problem. Hawking is a media superstar right up there with Einstein and Michael Jackson, so when reporters heard about this, the ensuing hoopla overshadowed everything else in the conference."

Can scientific truth be pursued in this atmosphere?

In any case, as to the physical picture beyond Hawking's technical lecture, his colleague and coworker Garry Gibbons says that[101]

- "Hawking's black holes, unlike the classical black holes, do not have a well defined event horizon that hides everything within them from the outside world.

In essence, his new black holes now never quite become the kind that gobble up everything. Instead, they keep emitting radiation for a long time, and eventually open up to reveal the information within."

But a BH is an absolute concept which involves formation of a $z = \infty$ surface which is the boundary of a trapped region from where nothing can escape. A BH would not "gobble up" everything only if $z = finite$, in other words if it is not a BH at all. Thus it is likely that the physical picture behind Hawking's turnaround could be that his realization that GC produces only high z surface and no EH. If it is really so, this would be the exactly what Mitra has shown. In the absence of a true BH (trivial topology?), there would not be any Information loss in the first place.

In one review article, Mario Rabinowitz[102], has pointed out that there may not be any "Hawking Radiation" at all. As mentioned by him, Hawking effect may be only a mathematical artifact because it violates a principle of QM:
which does not permit the physical particle wave function to have singularities at those space-time points at which the external field is regular and where there are no sources.

Almost similarly, we pointed out that, for a physical matter, the worldline cannot be lightlike at any regular region of spacetime, such as the EH, where there is no matter, no source of gravitation. And if the contrary appears to happen, that point of spacetime must be the singularity, consequently, area of the EH must be zero. However Rabinowitz could not go to this logical end because like any other QG worker he too is misled by the classical BH paradigm.

The basic incompatibility between the EH (unless the EH itself is the singularity) and QG has been brought out by t'Hooft[103]:

- "The singularity is not where the problem lies. It is the horizon. At first sight one expects that physics at the horizon essentially follows by considering local observers. At first sight, one appears to be able to derive all one wants to know from ordinary physics laws. Doing this, one finds that the horizon properties are locally determined.

But this does not work for the statistical interpretation of the entropy. One would expect that, whatever happens when gravitational collapse of some sort occurs, the state one enters into should have a finite entropy, and therefore represents a finite number of possible quantum states. But the properties of these quantum states do not at all follow from considerations concerning local observers. One always gets an infinite degeneracy of quantum states. It is this unphysical result that one wants to avoid, and it appears that entirely unorthodox physics is required."

He is absolutely right, the problem, indeed lies at the horizon. While these comments by t'Hooft are profound, we may recall that

• the occurrence of infinite energy, infinite Lorentz factor, infinite acceleration, infinite redshift at the EH cannot be transformed away. Such infinities at the EH cannot be explained "by ordinary physics laws". Also, there are host of other unphysical features associated with the EH even at the classical level.

• the occurrence of infinite quantum degeneracy (under the assumption of $M_0 > 0$) is directly related to the above mentioned infinities and which, in turn, is directly related to occurrence of singularities in the vacuum Hilbert metric at the EH.

• As per QM, there would not be any EH in the first place in order that $a \leq a_{Planck}$ and

• No unorthodox physics is really required to resolve this conundrum; on the other hand, all that is required is to appreciate the *unorthodox result* $\alpha_0 = 0$ and there is no point singularities in QM and GTR. In QG parlance, *it is the "back reaction" during GC which ensures that* $\alpha_0 = 2M_0 = 0$. More correctly QG will ensure, $M_0 \sim M_{Planck}$ and there is no EH at all. The "back reaction" to reach this stage need not be of QG origin, on the other hand, the classical H-K process would certainly generate continuous "back reaction".

Although, we have shown that the horizon and its predecessor, the trapped surface do not occur, there seems to be little excitement over this result at least at this moment.

17 Conclusion

The concept of BH grew over the past 90 years because

a) The integration constant ($\alpha = 2M$) appearing in the Hilbert solution for the space-time for a non-radiating spherically object of radius R_0 was considered to be finite even when $R_0 \to 0$. In other words, the Newtonian concept of a finite mass point singularity was, inadvertently and innocuously forced into GTR. Since our common sense and intuition is actually governed by Newtonian ideas, the choice of $\alpha_0 > 0$ seemed to be most natural.

b) The fact that the local Acceleration Invariant a diverges at $R = \alpha_0$ has been brought out rather recently and was not known earlier[3,12,13]. Divergence of a at the EH implies that the EH itself is the central physical singularity ($R_g = \alpha_0 = 2M_0 = 0$). This direct physical result, has, nonetheless, been highlighted by authors who do not have sufficiently prestigious affiliation or academic influence. Thus this crystal clear scientific result has been more or less ignored.

c) The important work of Stoner, Chandrasekhar, Landau, Oppenheimer and Volkoff and others firmly consolidated the idea of an upper mass limit of **cold** baryonic self-gravitating objects. At present this upper mass limit is broadly $M \sim (3-5)M_\odot$ assuming the central density of $\rho_c \sim 10^{14}$ g cm^{-3} and scales as ${\rho_c}^{-1/2}$. Since observationally, there are firm evidences of compact condensations with masses considerably or far above this limit it is believed that BHs have been detected. What is forgotten here is that the *BHCs need not be cold objects at all* and for which the just mentioned *mass limit is not applicable at all*.

d) Gravitational collapse problem has often been reduced to a tractable applied mathematics problem by dissociating physics from it. For instance, either pressure p has been set to zero and yet ρ has been assumed to be finite by ignoring both thermodynamics, hydrodynamics and causality. And even when a finite p has been considered, the associated H-K mechanism has been either completely ignored or practically ignored by making convenient assumptions. This has always given the false impression of formation of trapped surfaces and point singularities. In the resultant applied mathematics problem (where actually there should be no collapse if M is finite or fixed or else $M = 0$ if collapse is assumed) various assumptions about minor details have given rise to either Naked Singularities or BHs.

The idea that with diverging spacetime curvature near $R = 0$, the proper radial length too can diverge and the hence the collapse can proceed eternally (under classical GTR) has eluded everybody's (Newtonian) intuition.

e) The simple fact that QG gives a natural upper limit of proper acceleration a_P and hence there cannot be any EH (where $a = \infty$) too has eluded everybody's attention.

To recapitulate, one obtains the result $\alpha_0 = 0$ by interpreting the findings :

(i) divergence of a at $R = \alpha_0$ and

(ii) an originally timelike radial geodesic would become null at the same place.

17.1 Vaidya Metric and Non-occurrence of Trapped Surface and BH

The exterior spacetime of a collapsing and radiating body is described by the Vaidya metric:

$$ds^2 = \left(1 - \frac{2M(v)}{R_0}\right) dv^2 - 2dvdR_0 - R_0^2(d\theta^2 + \sin^2\theta d\phi^2) \tag{246}$$

where v is the retarded time. The determinant of this metric is

$$g = -R_0^4(1 - 2M/R_0)\ \sin^2\theta \tag{247}$$

The expression for g for the internal metric is not known. Yet both the metric and g must be continuous everywhere including the boundary of the body. Also, since now there is no vacuum spacetime, the question of the supposed "Schwarzschild Coordinate Singularity" also should not arise. Recall that for a supposed finite mass BH, one argued that the EH was a mere coordinate singularity because $g_H = -R^4 \sin^2\theta$ *appears* to be finite there. Thus we should be able to demand that $g \neq 0$ at the boundary always even if there would be any "coordinate singularity".

But Eq.(247) shows that $g = 0$ if $R_0 = 0$, i.e., if the body would collapse to a point mass, and this should, indeed be true. But *g also becomes zero if $R_0 = 2M$ or if an EH would form*. This fundamental result points out that there cannot be any EH. However, since the GR community, so far thought that $\alpha_0 = 2M_0 > 0$, it could not accept this fundamental result. On the other hand, with the premeditated idea of finite mass BHs, it again allowed itself to wallow in confusion/inconsistency: it tried to brush aside this fundamental result by ad hocly declaring that Vaidya metric would be valid only as long as EH is not formed!!

Note that, our result of non-occurrence of trapped surfaces $2M(r,t)/R \leq 1$ is inevitable in order that the Vaidya constraint on $2M/R_0$ gets matched with the interior result at $R = R_0$. In fact, Vaidya metric, on its own strength, tells that, the final state of gravitational collapse is

$$\frac{2M(v)}{R_0} = 1 \tag{248}$$

corresponding to $g = 0$, $g_{00} = 0$, and $z = \infty$. The collapse cannot progress any further because then one would have $g > 0$ in violation of GR! Since the final value of $R_0 = 0$, this equation also tells that, $M(R_0 = 0) = 0$, i.e., the final state of GR GC of sufficiently massive bodies is unique - It corresponds to a zero mass BH, which is borderline case between a BH and a Naked Singularity. Actually this singularity is approached only asymptotically and is never formed in any finite proper time because this corresponds to $ds^2 = 0$ for the material particles on the boundary of the fluid.

While the final state corresponds to $R_0 = 0$, what about the states having finite R_0? The same Vaidya metric suggests that the intermediate states must be radiating compact objects or ECOs rather than anything else. Why so? Because once the collapsing object crosses the Buchadal limit of $z = 2$, there cannot be any static state. If there is no static state there must be only, in a strict sense, collapsing and radiating states. Since $z \gg 1$, radiation is trapped within it and only a fraction $\sim (1+z)^{-2}$ can escape, and this meagre radiation is the Eddington luminosity. At the same time, at Eddington luminosity, an object is, for practical purpose static, though, in a strict sense, it is collapsing. And this is an ECO.

One would be tempted to know, if GTR is such a beautiful theory, can it not reveal this fundamental secret ($\alpha_0 \equiv 0$) in a more generous manner which would, at one go, clean up the accumulated 90 years of misconception?

We show below that GTR indeed is capable of doing so.

17.2 Invariance of 4-Volume Associated with a Point Mass

Let us recall the fundamental property associated with any curvilinear coordinate transformation[8,17,33,34]:

$$\int \sqrt{-g}\, d^4x = INVARIANT \tag{249}$$

and apply this to Eddington-Finkelstein and Hilbert coordinate systems associated with a BH

$$\int \sqrt{-g_*}\, dT_*\, dR\, d\theta\, d\phi = \int \sqrt{-g}\, dT\, dR\, d\theta\, d\phi \tag{250}$$

Note that since the null coordinates or Regge-Wheeler coordinates or Eddington- Finkelstein coordinates are obtained by integrating the *vacuum* metric all the way from $R = 0$, all such coordinate transformations can be associated only with BHs ($R_0 = 0$) and not with objects filled with mass-energy ($R_0 > 0$). Also, from Eq.(94), note that

$$-g_* = R^4 \sin^2\theta = -g \tag{251}$$

so that

$$\int R^2 \sin\theta \, dT_* \, dR \, d\theta \;\; d\phi = \int R^2 \sin\theta \, dT \, dR \, d\theta \;\; d\phi \tag{252}$$

Now carrying out the θ, ϕ integration on both sides of Eq.(251), we are left with

$$\int R^2 \, dT_* \, dR = \int R^2 \, dT \, dR \tag{253}$$

Recalling Eq.(93) and by substituting the relation

$$dT_* = dT + \frac{\alpha_0}{R - \alpha_0} dT \tag{254}$$

in Eq.(231) and by subtracting the common term from both sides we obtain

$$\alpha_0 \int \frac{R^2}{R - \alpha_0} \, dR \, dT = 0 \tag{255}$$

This result directly yields

$$\alpha_0 \equiv 0 \tag{256}$$

Thus the **integration constant α_0 which arose in the solution for the spacetime around a "massenpunkt", turned out, after 90 years, to be actually zero**. Its value, unfortunately, has, all along these 90 years, been considered to be finite because Newtonian massenpunkts have finite mass. Thus, actually, there is no spacetime beneath $R = 2M$ as per the correct intuition of Einstein and other founding fathers of GTR. Physically occurrence of $M_0 = 0$ for a massenpunkt implies its formation (in a gravitational collapse) would mathematically take infinite comoving proper time and, in reality, there is no "point particle" in GTR. Whereas a similar concept is used in the formulation of string theories of QG in a somewhat adhoc manner, this basic ingredient is naturally present in GTR, and this may unfurl a deeper connection between QM and GTR.

It can be also shown that there cannot be any spinning BHs either[104,105].

ECO has briefly been discussed (somewhat incorrectly) by Baryshev & Teerikorpi in "Discovery of Cosmic Fractals"[106]. MECO has found its place in one list of frequently used abbreviations used in Astronomy and Astrophysics[107]. Yet, a proper acceptance and appreciation of this fundamental fact may, unfortunately, take a long time in the existing all pervading all powerful paradigm of BHs which attempts to suppress any scientific critique pointing at it.

17.3 Was the Result not Known?

In a sense, the non-existence of finite mass BHs could have been announced way back in 1962 by Arnowitt, Deser and Misner! How?

In their classic paper (this author learnt of this paper only when these authors wisely decided to put it on net recently)[108], these authors found the mass of a spherical body by including its negative self-gravity by means of basic field theoretic approach. They found that for a static body of sufficiently small radius (i.e., R_0) ϵ, the gravitational mass is given by

$$M = G^{-1}\left(-\epsilon + [\epsilon^2 + 2M_b\epsilon G]^{1/2}\right) \tag{257}$$

where M_b is the "bare mass" of the body. Obviously $M > 0$ for $\epsilon > 0$. But as $\epsilon \to 0$, i.e. as one would think of a "Point Mass", $M \to M_0 = 0$. However, if the body has a charge e, the "point mass" will have a mass of purely electromagnetic origin[108]:

$$M = \frac{1}{2}(e^2/4\pi)G^{1/2} \tag{258}$$

So in the point mass limit, only the coupling of the electromagnetic self-field survives (through the presence of G). Thus for a neutral point mass, one has $M \to M_0 = 0$. If one thinks from the view point of GC, the entire body becomes a Point Mass ($\epsilon \to 0$) at the end of continued collapse. Thus end state of GC must correspond to $M_0 = 0$ (in the chargeless case) as found by us.

Alternately, even if one dissociates a BH from GC, then also, for an uncharged BH, the mass-energy stress tensor $T_{ik} = 0$ everywhere except at $R_0 = 0$. Thus mass of the BH must be entirely determined by the mass of the Massenpunkt at the central singularity. And this is zero as shown by Arnowitt, Deser and Misner[108].

Very significantly in footnote[20] of this classic paper, they also mentioned that

"It is interesting to note that even though $m = 0$ for the neutral particle, this does not imply that the space is flat everywhere flat for $r > \epsilon$ but rises steeply in the interior."

Here we may recall our Eq.(43). The value of the Kretschmann scalar on the EH is

$$K^{EH} = \frac{3}{4M_0^4} \tag{259}$$

And though $M_0 = 0$ for the massenpunkt or the BH, $K^{EH} = \infty$, i.e., it is a curvature singularity despite having zero mass. However for $R > R_0 = 0$, $K = 0$.

18 Dark Energy Stars ?

Although the idea of negative energy stars is probably old, it was practically forgotten by 2000. However after we showed that BHs are unreal, there has been a flurry of academic activity to find alternatives to BHs. As Cattoen, Tristan and Visser (CTV)[109] say

"Although the concept of a black hole is well-established and generally accepted in the relativity, astrophysics, and particle physics communities, one *sometimes encounters a certain amount of scepticism regarding the physical reality of the mathematical solution* , and wariness regarding the interpretation of observational data." (Emphsasis by the author).

Of course, CTV do not mention that this *scepticism* got strengthened after Mitra showed that there cannot be finite mass BHs and Robertson & Leiter showed that the so-called BHCs have strong intrinsic magnetic field which is an evidence against the EH.

The paper of CTV, nevertheless, is quite important because it has shown that all the "Gravstar" and "Dark Energy Star" models involving isotropic negative pressure are completely untenable (even if one would accept "negative pressure" and exotic phase transitions) because

• At the boundary, GR equation for hydrostatic equation *cannot be satisfied with negative interior pressure*

• They actually *contain pressure singularity or curvature singularity*

CTV therefore have attempted to have Gravstars with anisotropic pressure. But in truly spherical symmetry, a static fluid cannot be anisotropic. Anisotropy might appear if there is a the global magnetic field (actually, a global magnetic field would, in a strict sense, spoil, spherical symmetry). The only way pressure anisotropy could appear despite a strict spherical symmetry if there is any flow (like heat or the fluid itself) in the radial direction. In such a case, despite spherical symmetry, a "preferred direction" of heat flow may be conceived and there could be attendant pressure anisotropy. But in the *cold* and *static* case, there cannot be any pressure anisotropy. However, if instead of a fluid, one considers an unrealistic model of "dusts" there could be only tangential pressure (due to rotation of dust particles) while radial pressure would be zero.

Since a "Gravstar" or any other body is no collection of rotating dust particles, there cannot be any meaningful anisotropic Dark Energy Star.

Further all such *cold* and *exotic* bodies would not have any intrinsic magnetic field. And they should exhibit Type I x-ray bursts by virtue of their unmagnetized "hard surfaces". But the BHCs do not show any Type I burst activity[110]. On the other hand, ECOs/MECOs will not show any Type I X-ray burst activity because they have strong intrinsic magnetic field (the reason that x-ray pulsar Her X-1 does not entertain any Type I x-ray burst). Further the surface of a ECO/MECO would be extremely hot gaseous type rather than "hard".

Similarly, though a ECO/MECO has a physical surface and magnetic field there will not be any spin pulsation from it because of extremely strong GR frame dragging effect[111].

Thus the only meaningful alternative to BHs is an ECO or its ultramagnetized version MECO. This is so because of the extreme bending of photon and neutrino orbits at high z, the trapped radiation pressure will grow as $\sim (1+z)^2$ and the collapse would inevitably come to a dynamic halt at a finite z.

18.1 Epilogue

As to the rejuvenated interest on Quantum Information Loss Paradox in 2004, specifically, whether "Information" comes out of BH or not, Paul Ginsparg mentions that there was a "voting". While most voted for (i) information was really lost and many voted for (ii) information comes out, a minority voted for an unspecified (iii) "something else".

The explicit proofs that $\alpha_0 \equiv 0$ indeed confirmed that it is indeed "something else". There was no BH to begin with, the continued collapse indeed continues indefinitely in the form of ECOs which have no horizon. So indeed "*The Truth is out there*. And this truth not only pertains to

(i) Quantum Information Paradox; but on the other hand it pertains to one of the oldest and most fundamental query of Physical Sciences, namely, (ii)"What is the final State of Gravitational Collapse?" - the answer is either cold objects like White Dwarfs, Neutron Stars or hot objects like ECO whose asymptotic final state is a zero mass BH, a borderline case between the notions of a BH and a Naked Singularity.

The truth is out there for the X-ray astronomers and High Energy Astrophysicists as well; all those compact objects much more massive than cold Neutron Stars, whether they occur in the galaxy or anywhere are ECOs/MECOs, they have physical surface, spin and intrinsic magnetic field and the combination of which often produce high energy emissions of the cosmos[113] and jets at all scales.

But are we ready to appreciate this "Truth" unmindful of our prejudices and personal interests?

Acknowledgement

Without the academic and moral support provided by Prof. P.C. Vaidya, the most senior and most respected relativist of India, in 1999, such a review may not have been written now. When Physical Rev. Lett. rejected both my initial papers 57, 58] without showing any valid reason, Prof. Vaidya suggested that he would forward them to Indian Journal "Current Science" with his recommendations. A word of encouragement from my M.Sc. teacher Prof. A.K. Raychowdhury (who expired recently), another doyen of Indian relativity research, was also helpful. I am also indebted to my friend Sabbir Ahmed, a person capable of both appreciating and having original thinking, for constant encouragement and great help. Over these years I have greatly benefitted from regular academic exchanges with Stanley Robertson and Darryl Leiter. I thank Pankaj Joshi for sparing his time for several detail academic sessions and other helps; I appreciate his existential dilemma in offering correct comment on my work in public.

When I submitted a relevant paper to Foundations of Physics in 1999, there was no response, whatsoever, from its editor for almost 3 months. Then I contacted Prof. G. t. Hooft as he was on the Editorial Board of this journal. It was only after his gentle prod, that the Editor, Prof. Alwyn van Merwe, resumed the processing of the manuscript with vigour. I am grateful to Prof. Hooft for this help and naturally I was very happy when he won the Nobel in the following year. But I am also grateful to Prof. Merwe for his great patience, fairness, professionalism and courage shown during the processing of this manuscript.

Ever since I landed up on this piece of GR GC research, I came under tremendous pressure, spanning from threat to ridicule, from various directions including two of my superiors. Yet several individuals/friends gave me great encouragement which enabled me to carry on. I would like to thank some of them:

Harish Parab, Arun Pati, Sudhir Jain, S. Ganeshan, Zafar Ahmed, S. Acharya, Hiren Bose, Subrata Bose, Miloni Bhat, Komal Mehta, K. Jayaraman, Lalitha Vaidyanathan and many others including one Radio BBC correspondent whose name I have forgot now. Finally I thank Paul Ginsparg, the visionary behind Los Alamos Preprint Archive.

Eventually, this work is dedicated to the event of formulation of Theory of Relativity in 1905,i.e., 100 years ago.

References

[1] P.S. Joshi, Gravitational Collapse End States (2004) (gr-qc/0412082)

[2] A. Mitra, Non-occurrence of trapped surfaces and black holes in spherical collapse; An abridged version, *Found. Phys. Lett.* **13**(6), 543 (2000) (astro-ph/9910408)

[3] A. Mitra, On the final state of spherical gravitational collapse, *Found. Phys. Lett.* **15**(5), 439 (2002) (astro-ph/0207056)

[4] A. Mitra, On the nature of the compact condensations at the center of galaxies, *Bull. Astr. Soc. India*, **30**, 173 (2001) (astro-ph/0205261).

[5] A. Mitra, On the question of trapped surfaces and black holes (2001), Invited Talk in Black Hole Workshop organized by Center for Space Physics, Kolkata, (2001) (astro-ph/0105532)

[6] S. Chandrasekhar, *An Introduction to the Study of Stellar Structure* (Dover, New York, 1967)

[7] J. Michell, *Phil. Trans.*, **74**, 35 (1784)

[8] L.D. Landau, & E.M. Lifshitz, *The Classical Theory of Fields* (Pergamon Press, Oxford, 1962)

[9] L.D. Landau & E.M. Lifshitz, *The Classical Theory of Fields*, 4th ed., (Pergamon Press, Oxford, 1985).

[10] Hilbert, D., Nachr, Ges. Wiss. Gottingen, *Math. Phys. Kl.*, **53** (1917)

[11] K. Schwarzschild, *Sitzungsberichte Preuss. Acad. Wiss.*, **424** (1916), for English translation, see, S. Antoci and A. Loinger, (1999) physics/9905030 and

[12] L.S. Abrams, *Can. J. Phys.*, **67**, 919 (1989) (gr-qc/0102055)

[13] S. Antoci & D.-E. Liebscher (2004) (gr-qc/0406090)

[14] S. Antoci and D.E. Liebscher, (2001) (gr-qc/0102084) also see A. Loinger, (2000), (astro-ph/0001453), A. Loinger, (1999) (gr-qc/9908009), and Z. Zakir, (1999) (gr-qc/9905068)

[15] G.D. Birchoff, *Relativity and Modern Physics*, (Harvard Univ. Press, Cambridge, Mass. 1923)

[16] P. Townsend, *Black Holes: Lecture Notes* (1997) (gr-qc/9707012)

[17] C.W. Misner, K.S. Thorne, and J.A. Wheeler, *Gravitation*, (W.H. Freeman and Company, San Francisco, 1973)

[18] R.H. Fowler, *Mon. Not. Roy. Astr. Soc.*, **87**, 114, (1926)

[19] W. Anderson, *Zs.f. Phys.*, **54**, 433 (1929)

[20] E.C. Stoner, *Phil. Mag.*, **7**, 63, (1929)

[21] E.C. Stoner, *Phil. Mag.*, **9**, 944, (1930)

[22] S. Chandrasekhar, *Mon. Not. Roy. Astr. Soc.*, **91**, 456 (1931) and *Astrophys. J.*, **74**, 81, (1931)

[23] S. Chandrasekhar, *Mon. Not. Roy. Astr. Soc.*, **95**, 207, (1935)

[24] R.F. Tooper, *Astrophys. J.*, **140**, 349 (1964)

[25] J.R. Oppenheimer & G.M. Volkoff, *Phy. Rev.*, **55**, 374 (1939)

[26] S.L. Shapiro & S.A. Teukolsky, *Black Holes, White Dwarfs and Neutron Stars: The Physics of Compact Objects*, (Wiley, New York, 1983)

[27] L.D. Landau, *Phys. Z. Swjetunion*, **1**, 285, (1932).

[28] J.R. Oppenheimer and H. Snyder, *Phys. Rev.* **56**, 455 (1939)

[29] A. Mitra, The Mass of the Oppenheimer-Snyder Black Hole (1999) (astro-ph/9904163)

[30] A. Einstein, *Ann. Math.* **40**, 922 (1939)

[31] P.O. Mazur & E. Mottola, Gravitational Condensate Stars: An Alternative to Black Holes (gr-qc/0109035)

[32] D. Leiter & S. Robertson, Does Principle of Equivalence prevent trapped surfaces from being formed in general relativistic collapse process, *Foun. Phys. Lett.* **16**, 143 (2003) (astro-ph/0111421).

[33] H. Stephani, *General Relativity* (Cambridge Univ. Press, Cambridge, 1990)

[34] B.F. Schutz, *A First Course in General Relativity*, (Cambridge Univ. Press, Cambridge, 1986)

[35] N. Rosen, in *Relativity*, eds. M. Carmeli, S.I. Fickler, & L. Witten (Plenum, New York, 1970)

[36] A.S. Eddington, *Nature*, **113**, 192 (1924)

[37] D. Finkelstein, *Phys. Rev.*, **110**, 965 (1958)

[38] G. Lemaitre, *Ann. Soc. Sci. Bruxelles*, **53**, 51, (1933)

[39] M.D. Kruskal, *Phys. Rev.*, **119**, 1743 (1960)

[40] G. Szekeres, *Publ. Math. Debrecen*, **7**, 285 (1960)

[41] A. Mitra, Kruskal Dynamics for Radial Geodesics.I (1999) (gr-qc/9909062) also see R. Burghardt, The Onefold Truth (physics/0506045)

[42] A.A. Logunov, M.A. Mestverishvili & V.V. Kiselev, Black Holes: a prediction of theory or fantasy? (2004) (gr-qc/0412058)

[43] R. Wald, *General Relativity*, (Chicago Univ. Press, Chicago, 1984)

[44] S. Weinberg, *Gravitation and Cosmology*, (John Wiley, New York, 1972)

[45] S. Robertson, Private Communication (2003)

[46] S.W. Hawking & G.F.R. Ellis, *The Large Scale Structure of Space-time*, (Cambridge Univ. Press, Cambridge, 1973).

[47] P. Bizon, E. Malec, & N. O'Murchadha, Trapped Surfaces in Spherical Stars, *Phys. Rev. Lett.* **61**(10), 1147 (1988)

[48] N. Dadhich, Nariai metric is the first example of the singularity free model (2003) (gr-qc/0106023)

[49] H. Nariai, *Sci. Rep. Tohoku Univ.* **34**, 160 (1950)

[50] A. Banerjee, S. Chatterjee, & N. Dadhich, *Mod. Phys. Lett. A* **17**, 2335 (2002) (gr-qc/0209035)

[51] J.M.M. Senovilla, *Phys. Rev. Lett.* **64**, 2219 (1990).

[52] S.M. Goncalves, Absence of trapped surfaces and singularities in cylindrical collapse, *Phys. Rev. D* **65**, 4045G (2002)

[53] A.Mitra, A New Proof for Non-occurrence of Trapped Surfaces and Information Paradox (2004) (astro-ph/0408323)

[54] C.W. Misner and D.H. Sharp, *Phys. Rev.* **136** 2B, 571 (1964).

[55] M. May and R. White, Phys. Rev. B **141**, 1233 (1965).

[56] W.C. Hernandez and C.W. Misner, *Astrophys. J.* **143**, 452 (1966); also see, R.W. Lindquist, R.A. Schwartz, and C.W. Misner, *Phys. Rev. B* **137** , 1364 (1965); C.W. Misner, *Phys. Rev.* **137** , B 1360 (1965)

[57] A. Mitra, Towards the Final State of Spherical Gravitational Collapse and Likely Source of Gamma Ray Bursts (1998) (astro-ph/9803013)

[58] A. Mitra, Final State of Spherical Gravitational Collapse and Likely Source of Gamma Ray Bursars (1998) (astro-ph/9803014)

[59] A. Mitra, Non-occurrence of Trapped Surfaces and Black Holes in Spherical Gravitational Collapse (1998) (gr-qc/9810038)

[60] A. Mezzacappa, Talk in "Supernovae as Cosmological Lighthouses" (Padua, Italy, 2004) (astro-ph/0410085)

[61] A. Mahajan, R. Goswami & P.S. Joshi, *Class. Quant. Grav.*, **22**, 271 (2004) (gr-qc/0312015)

[62] R. Goswami & P.S. Joshi, *Class. Quant. Grav.*, **21**, 3645 (2004) (gr-qc/0406052)

[63] R. Goswami & P.S. Joshi, *Phys. Rev. D*, **69**, 027502 (gr-qc/ 0310122)

[64] R. Chan et al., Friedmann Like Collapse of Radiating Spheres, *ApJ*, **342**, 976 (1989)

[65] F.I. Cooperstock, S. Jhingan, P.S. Joshi, & T.P. Singh, *Class. Quan. Grav.*. **14**, 2197 (1997)

[66] T. Harada, *Pramana*, **53**(6), 1, (1999), (gr-qc/0407107)

[67] H.J. Mosquera Cuesta, J.M. Salim, & N.O. Santos, **Gravitationally Eternally Collapsing Objects**, Paper presented in "100 years of Relativity, Sao Paula, Brazil, 2005)

(fma.if.usp.br/100years/Abstracts.htm , www.sab-astro.org.br/sab31/comunicacoes-orais.pdf)

[68] A. Mitra, Magnetospheric Eternally Collapsing Objects (MECOs): Likely New Class of Source of Cosmic Particle Acceleration, *Proc. 29th Int. Cos. Ray Conf.*, (Pune 2005) (physics/0506183)

[69] On the Probable Nature of the "Dark Particle Accelerators" Discovered by HESS, *Proc. 29th Int. Cos. Ray Conf.*, (Pune 2005) (astro-ph/0510162)

[70] R. Penrose, *Nouvo Cimento*, **1**, 533 (1965)

[71] S. Robertson, & D. Leiter, Evidence for intrinsic magnetic moment in black hole candidates, *Astrophys. J.* **565**, 447 (2002) (astro-ph/0102381).

[72] S. Robertson, & D. Leiter, On the intrinsic magnetic moment in black hole candidates *Astrophys. J.* **569**, L203 (2003) (astro-ph/0310078)

[73] S. Robertson, & D. Leiter, On the origin of the radio/X-ray luminosity correlation in black hole candidates, *Mon. Not. Roy. Astr. Soc.* **350**, 1391 (2004) (astro-ph/0402445)

[74] S. Robertson & D. Leiter, How Black are Black Hole Candidates (2002) (astro-ph/0208333)

[75] S. Robertson & D. Leiter, Do Black Hole Candidates Have Magnetic Moments Instead of Event Horizons? (2003) (astro-ph/0307438)

[76] E. Malec, & N. O'Murchadha, *Class. Quant. Grav.*, **21**, 5777, (2004) (gr-qc/0408044)

[77] Y. Tomozawa, (2004) (gr-qc/0405071)

[78] J. Ehlers, Private Communication (2000)

[79] S. Chakarabarti, *Mon. Not. Roy. Astron. Soc.*, **283**, 325, (1996); also Private Communication (2001)

[80] R. Schoen & S-T Yau, Positivity of the Total Mass of a General Space-Time, *Phys. Rev. Lett.*, **43**, 1457, (1979), some more such examples are,

R. Bartnik, New Definition of Quasilocal Mass, *Phys. Rev. Lett.*, **62**, 2346, (1989),

C-C. M. Liu & S-T Yau, Positivity of Quasilocal Mass, *Phys. Rev. Lett.*, **90**, 231102-1 (2003)

[81] S.W. Hawking, *Commun. Math. Phys.*, **43**, 199 (1975)

[82] T. Padmanabhan, Private Communication (1998)

[83] D. Lohiya, Comments made during a Chat Show in *Headlines Today*, (Indian TV News Channel) (Aug.6, 2004), also see his comments in a Blog site "My Messy Workbench", posted from E.ID. slohiya@indiatimes.com

[84] A. Ashtekar, Private Communication (1999)

[85] P. Bizon, E. Malec, & N.O'. Murchadha, *Class. Quant. Grav.*, **7**, 1953, (1990)

[86] E. Sullivan, Editorial Reply after 1st round of review of Ref.[57] in *Annals of Physics*, (1999)

[87] J. Baez, www.sci.physics.research (1999-2004) (for the threads, search with "Mitra"or search *google groups*)

[88] C. Hillman, www.sci.physics.research (1999-2004) (for the threads, search with "Mitra" or search *google groups*), http://www.math.washington.edu/ hillman/

[89] J. Baez, http://math.ucr.edu/home/baez/RelWWW/wrong.html

[90] J. Baez in www.sci.physics.relativity (search with "Mitra")

[91] J. Baez in www.sci.physics.relativity (search with "Mitra")

[92] J. Baez, http://math.ucr.edu/home/baez/RelWWW/history.html

[93] G. Horowitz, The dark side of string theory: Black Holes and Black Strings (1992) (hep-th/9210119)

[94] A. Strominger, Massless Black Holes and Conifolds in String Theory, *Nucl. Phys. B*, **451**, 109 (1995) (hep-th/954090)

[95] A. Sen, Extremal Black Holes and Elementary String States, *Mod. Phys. Lett. A*, **10**, 2081 (1995) (hep-th/9504147)

[96] S. Mathur, Where are the states of a black hole? (2003) (hep-th/0401115)

[97] A.K. Pati, *Euro. Phys. Lett.*, **18**, 285 (1992), also *IL. Nuovo Cimento B*, **107**, 895 (1992)

[98] see www.sa.infn.it/dipsa/index.html

[99] S.W. Hawking (2004), see GR17 homepage, http://www.dcu.ie/ nolanb/gr17.htm

[100] J. Baez, (2004) see http://math.ucr.edu/home/baez/week207.html

[101] Jenny Hogan, "Hawking cracks black hole paradox" in *NewScientist.com* 14th July (2004)

[102] M. Rabinowitz, Black Hole Paradoxes, (astro-ph/0412101)

[103] G. t' Hooft, Private (email) Commun. (1998)

[104] A. Mitra, Why the astrophysical Black Hole Candidates are not rotating black holes, (2004) (astro-ph/0407501)

[105] A. Mitra, Why the astrophysical Black Hole Candidates may not be black holes at all, (2004) (astro-ph/0409049)

[106] Y. Baryshev & P. Teerikorpi, Discovery of Cosmic Fractals, (World Scientific, 1993).

[107] List of Frequently Used Abbreviations in A & A: www.phys.uni-sofia.bg/ astro/abbrev.html

[108] R. Arnowitt, S. Deser & C.W. Misner, The Dynamics of General Relativity in *Gravitation: An Introduction to Current Research* (ed. L. Witten, Wiley, NY, 1962) (gr-qc/0405109)

[109] C. Cattoen, T. Faber, & M. Visser, Gravstars must have anisotropic pressures, (gr-qc/0505137), also see

A. Mitra, Comments on the proposal of Dark Energy Stars by Chapline, (astro-ph/0504384)

[110] A. Mitra, On the non-occurrence of Type I X-ray Bursts from the Black Hole Candidates, *Proc. COSPAR Colloquium on Spectra and Timing of Compact X-ray Binaries* , (Mumbai, 2005) (astro-ph/0510162)

[111] A. Mitra, Why the observed Black Hole Candidates do not show any spin pulsation, *Proc. COSPAR Colloquium on Spectra and Timing of Compact X-ray Binaries* , (Mumbai, 2005) (astro-ph/0510205)

[112] P. Ginsparg, The Truth is Still Out There (physics/0408033), also

www.nytimes.com/2004/08/03/opinion/03ginsparg.html

[113] S. Robertson & D. Leiter, Magnetospheric Eternally Collapsing Objects (MECO) Model of Galactic Black Hole Candidates and Active Galactic Nuclei, Invited Review article in "Focus on Black Hole Research" (Nova, NY, 2005, in press).

In: Focus on Black Hole Research
Editor: Paul V. Kreitler, pp. 95-104
ISBN 1-59454-460-3

Chapter 2

BLACK HOLES AND MASSIVE ELEMENTARY PARTICLES IN RESUMMED QUANTUM GRAVITY*

B.F.L. Ward
Department of Physics,
Baylor University, Waco, Texas 76798-7316, USA

Abstract

Einstein's general theory of relativity poses many problems to the quantum theory of point particle fields. Among them is the fate of a massive point particle. Since its rest mass exists entirely within its Schwarzschild radius, in the classical solutions of Einstein's theory, the respective system should be a black hole. We address this issue using exact results in a new approach to quantum gravity based upon well-tested resummation methods in point particle quantum field theory. We show that the classical conclusion is obviated by quantum loop effects. We show that our new approach already passes two theoretical checks with the published literature; for, it reproduces known results on the one-loop correction to the graviton self-energy in scaler matter coupled to Einstein's gravity as analyzed by 't Hooft and Veltman and it is consistent with the asymptotic safety results of Bonanno and Reuter on the behavior of Newton's constant in the deep Euclidean regime. Indeed, our approach is consistent with the black hole phenomenology of the latter authors, including their results on the final state of the Hawking radiation for an originally massive black hole. Further black hole related phenomenological implications are also discussed.

1 Introduction

Einstein's general theory of relativity has had many successful tests [1,2]. One of its most direct classical predictions is that a massive elementary point particle should be a black hole solution; for, in such a system, the rest mass lies entirely within the Schwarzschild

*Work partly supported by NATO Grant PST.CLG.980342.

radius. For the Standard Model [3, 4] such a conclusion would be at best problematic, as many of the elementary particles in the SM have non-zero rest mass and yet they are able to communicate themselves entirely in the well-tested SM point-particle field theoretic interactions – they do not only interact with the world beyond their Schwarzschild radii by Hawking radiation [5]. A key point then is whether or not quantum loop effects obviate the classical general relativistic conclusion that a massive point particle is a black hole. It is this question that we answer in the following.

We base our analysis on pioneering work by Feynman [6, 7] on the quantum theory of general relativity in which he argued that Einstein's theory is just another point particle field theory in which the metric field of space-time undergoes quantum fluctuations as do all the other point particle fields in the SM. Feynman worked-out the Feynman rules for the simplest case of a massive scalar field coupled to Einstein's gravity and we will use his results in what follows. His formulation of quantum general relativity however was badly behaved in the deep Euclidean (ultra-violet(UV)) regime and it can be characterized as being non-renormalizable [8, 9]. What we do pedagogically beyond what is done in Refs. [6, 7] is to re-arrange *exactly* the Feynman series for quantum general relativity derived therein using our recent extension of YFS [10, 11] resummation methods to non-Abelian gauge theories [12]. The resummed theory which we derive [13], which will be seen to have improved UV behavior, is therefore not an approximation. We call it resummed quantum gravity (RQG). Since we do not modify the theory of Einstein or the quantum mechanics, we arrive at a minimal union of the ideas of Bohr and Einstein, in contrast to the popular superstring theory [14, 15] or the recently developing loop quantum gravity theory [16](LQG), where the existence of the smallest length parameter, the Planck length, means that at least Einstein's theory is modified in these two approaches to finite quantum loop effects in quantum general relativity[1].

The basic physical idea behind our resummation efforts is the following. In the over-all space-time point of view of Feynman, a point particle of mass m at a point x and another such particle at y experience an attractive force $\propto m^2$ due to Newton's law. When this is considered in the deep Euclidean regime, wherein the effective value of m^2 is large and negative, it translates into a large repulsive interaction against the propagation of the particle between the respective x and y. One concludes that the propagation of the particle should be severely damped in the deep Euclidean regime in the exact solutions of quantum general relativity. This suggests that we resume the theory to get better behavior of the Feynman series in the deep Euclidean regime.

We point-out that, in Ref. [18], Weinberg has noted the four approaches to the bad UV behavior of quantum general relativity:

- extended theories of gravitation: supersymmetric theories - superstrings; loop quantum gravity, etc.

- resummation

[1]In the superstring theory, it appears that quantum mechanics is also modified [17].

- composite gravitons
- asymptotic safety: fixed point theory (see Ref. [19,20])

Here, we are developing a new version of the resummation approach. We will make contact with the more phenomenological asymptotic safety approach as realized in Refs. [19,20] especially as it relates to black hole physics. Some speculations about the possible relationship between our new RQG theory and the superstring theory can be found in Ref. [13,21], wherein we also point-out the complementarity between our analysis and the large distance analyses in Ref. [22].

More precisely, after reviewing the formulation of Einstein's theory by Feynman in the next Section, we use YFS resummation methods in Section 3 to arrive at a theory of quantum general relativity in which the loop corrections are finite. In Section 4, we use these finite loop corrections to address the fate of massive elementary particles from the standpoint of their being black holes. We also show in this Section how our results for the UV behavior of our quantum loop effects relate to the corresponding results of the asymptotic safety approach in Refs. [19,20]. We argue that this relationship leads us to the same black hole physics phenomenology as that found in Refs. [19,20]. Section 5 contains some summary remarks.

2 Point Particle Field Theoretic Formulation of Einstein's Theory

In Feynman's point particle formulation [6, 7] of Einstein's theory we start with the Lagrangian density of the currently observed world

$$\mathcal{L}(x) = -\frac{1}{2\kappa^2}\sqrt{-g}R + \sqrt{-g}L^{\mathcal{G}}_{SM}(x) \tag{1}$$

where R is the curvature scalar, $-g$ is the negative of the determinant of the metric of space-time $g_{\mu\nu}$, $\kappa = \sqrt{8\pi G_N} \equiv \sqrt{8\pi/M_{Pl}^2}$, where G_N is Newton's constant, and the SM Lagrangian density, which is well-known (see for example, Ref. [3, 4, 23]) when invariance under local Poincare symmetry is not required, is here represented by $L^{\mathcal{G}}_{SM}(x)$ which is readily obtained from the familiar SM Lagrangian density as described in Refs. [13,21] as follows: since $\partial_\mu\phi(x)$ is already generally covariant for any scalar field ϕ and since the only derivatives of the vector fields in the SM Lagrangian density occur in their curls, $\partial_\mu A^J_\nu(x) - \partial_\nu A^J_\mu(x)$, which are also already generally covariant, we only need to give a rule for making the fermionic terms in usual SM Lagrangian density generally covariant. For this, we introduce a differentiable structure with $\{\xi^a(x)\}$ as locally inertial coordinates and an attendant vierbein field $e^a_\mu \equiv \partial\xi^a/\partial x^\mu$ with indices that carry the vector representation for the flat locally inertial space, a, and for the manifold of space-time, μ, with the identification of the space-time base manifold metric as $g_{\mu\nu} = e^a_\mu e_{a\nu}$ where the flat locally inertial space indices are to be raised and lowered with Minkowski's metric η_{ab} as usual.

Associating the usual Dirac gamma matrices $\{\gamma_a\}$ with the flat locally inertial space at x, we define base manifold Dirac gamma matrices by $\Gamma_\mu(x) = e^a_\mu(x)\gamma_a$. Then the spin connection, $\omega^a_{\mu b} = -\frac{1}{2}e^{a\nu}\left(\partial_\mu e^b_\nu - \partial_\nu e^b_\mu\right) + \frac{1}{2}e^{b\nu}\left(\partial_\mu e^a_\nu - \partial_\nu e^b_\mu\right) + \frac{1}{2}e^{a\rho}e^{b\sigma}\left(\partial_\rho e_{c\sigma} - \partial_\sigma e_{c\rho}\right)e^c_\mu$ when there is no torsion, allows us to identify the generally covariant Dirac operator for the SM fields by the substitution $i\not\partial \to i\Gamma(x)^\mu\left(\partial_\mu + \frac{1}{2}\omega_{\mu b}{}^a\Sigma^b{}_a\right)$, where we have $\Sigma^b{}_a = \frac{1}{4}\left[\gamma^b, \gamma_a\right]$ everywhere in the SM Lagrangian density. This will generate $L^{\mathcal{G}}_{SM}(x)$ from the usual SM Lagrangian density $L_{SM}(x)$ as it is given in Refs. [3,4,23], for example.

The fundamental issue we address here is the fate of the massive point particles in the SM. We do not expect the respective spin representation to be crucial to the conclusions we reach so we follow Feynman and replace (1) with the simplest example of our problem, the Lagrangian of a massive scalar field coupled to Einstein's gravity, where we have in mind the physical Higgs field of the SM, whose rest mass is known to be greater than 114 GeV with a 95% CL [24]:

$$\begin{aligned}
\mathcal{L}(x) &= -\frac{1}{2\kappa^2}R\sqrt{-g} + \frac{1}{2}\left(g^{\mu\nu}\partial_\mu\varphi\partial_\nu\varphi - m_o^2\varphi^2\right)\sqrt{-g} \\
&= \frac{1}{2}\left\{h^{\mu\nu,\lambda}\bar{h}_{\mu\nu,\lambda} - 2\eta^{\mu\mu'}\eta^{\lambda\lambda'}\bar{h}_{\mu_\lambda,\lambda'}\eta^{\sigma\sigma'}\bar{h}_{\mu'\sigma,\sigma'}\right\} \\
&\quad + \frac{1}{2}\left\{\varphi_{,\mu}\varphi^{,\mu} - m_o^2\varphi^2\right\} - \kappa h^{\mu\nu}\left[\overline{\varphi_{,\mu}\varphi_{,\nu}} + \frac{1}{2}m_o^2\varphi^2\eta_{\mu\nu}\right] \\
&\quad - \kappa^2\left[\frac{1}{2}h_{\lambda\rho}\bar{h}^{\rho\lambda}\left(\varphi_{,\mu}\varphi^{,\mu} - m_o^2\varphi^2\right) - 2\eta_{\rho\rho'}h^{\mu\rho}\bar{h}^{\rho'\nu}\varphi_{,\mu}\varphi_{,\nu}\right] + \cdots
\end{aligned} \tag{2}$$

Here, $\varphi(x)_{,\mu} \equiv \partial_\mu\varphi(x)$, and $g_{\mu\nu}(x) = \eta_{\mu\nu} + 2\kappa h_{\mu\nu}(x)$ where we follow Feynman and expand about Minkowski space so that $\eta_{\mu\nu} = diag\{1, -1, -1, -1\}$. Following Feynman, we have introduced the notation $\bar{y}_{\mu\nu} \equiv \frac{1}{2}\left(y_{\mu\nu} + y_{\nu\mu} - \eta_{\mu\nu}y_\rho{}^\rho\right)$ for any tensor $y_{\mu\nu}$[2]. Thus, m_o is the bare mass of our free Higgs field and we set the small tentatively observed [25] value of the cosmological constant to zero so that our quantum graviton has zero rest mass. The Feynman rules for (2) have been essentially worked out by Feynman [6, 7], including the rule for the famous Feynman-Faddeev-Popov [6, 26] ghost contribution that must be added to it to achieve a unitary theory with the fixing of the gauge (we use the gauge of Feynman in Ref. [6], $\partial^\mu\bar{h}_{\nu\mu} = 0$), so we do not repeat this material here.

We wish to use the quantum loop corrections predicted by (2) to study the black hole character of our massive elementary particle φ. The original Feynman series derived in Refs. [6, 7] is badly UV divergent. For example, the graphs in Fig. 1 are superficially quarticly divergent and, since they give us the scalar one-loop contribution to the graviton propagator, they will lead to a logarithmic UV divergence in the coefficient of q^4 in the respective 1PI 2-point function, a divergence which can not be removed by any amount of field and mass renormalization, i.e., these graphs already exhibit the non-renormalizability of the original Feynman series derived in Ref. [6,7].

In the next Section, with an eye toward our black hole physics analysis, we improve on it here by using the extension of the methods of Yennie, Frautschi and Suura [10] to

[2]Our conventions for raising and lowering indices in the second line of (2) are the same as those in Ref. [7].

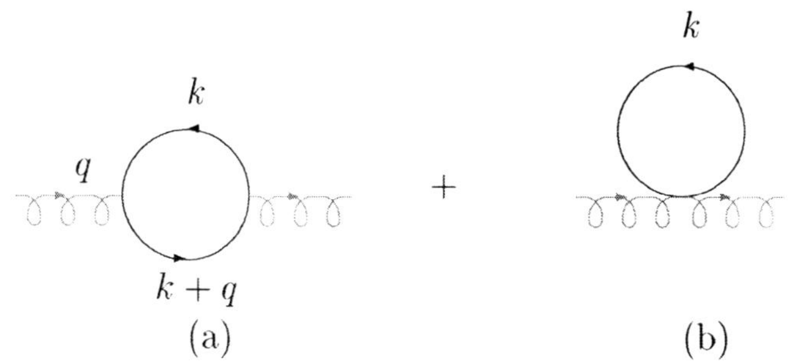

Figure 1: The scalar one-loop contribution to the graviton propagator. q is the 4-momentum of the graviton.

non-Abelian gauge theories as we have developed this extension in Ref. [12].

3 Resummed Quantum Gravity

In this section, we will YFS resume the propagators in the theory in (2): from the YFS formula (see eq.(5.18) in Ref. [10])

$$iS'_F(p) = \frac{ie^{-\alpha B''_\gamma}}{S_F^{-1}(p) - \Sigma'_F(p)}, \tag{3}$$

where $\Sigma'_F(p)$ is the sum of the YFS loop residuals, we need to find for quantum gravity the analogue of

$$\alpha B''_\gamma = \int d^4\ell \frac{S''(k,k,\ell)}{\ell^2 - \lambda^2 + i\epsilon} \tag{4}$$

where λ is the IR cut-off and

$$S''(k,k,\ell) = \frac{-i8\alpha}{(2\pi)^3} \frac{kk'}{(\ell^2 - 2\ell k + \Delta + i\epsilon)} \frac{1}{(\ell^2 - 2\ell k' + \Delta' + i\epsilon)}\Big|_{k=k'}, \tag{5}$$

for $\Delta = k^2 - m^2$, $\Delta' = k'^2 - m^2$. To this end, note also

$$\begin{aligned} \alpha B''_\gamma &= \int \frac{d^4\ell}{(2\pi)^4} \frac{-i\eta^{\mu\nu}}{(\ell^2-\lambda^2+i\epsilon)} \frac{-ie(2ik_\mu)}{(\ell^2-2\ell k+\Delta+i\epsilon)} \\ &\quad \frac{-ie(2ik'_\nu)}{(\ell^2-2\ell k'+\Delta'+i\epsilon)}\Big|_{k=k'}, \end{aligned} \tag{6}$$

where $\Delta = k^2 - m^2$, $\Delta' = k'^2 - m^2$ and λ is the IR cut-off. With the identifications [9] of the conserved graviton charges via $e \to \kappa k_\rho$ for soft emission from k we get the analogue ,$-B''_g(k)$, of $\alpha B''_\gamma$ by replacing the γ propagator in (6) by the graviton propagator,

$$\frac{i\frac{1}{2}(\eta^{\mu\nu}\eta^{\bar\mu\bar\nu} + \eta^{\mu\bar\nu}\eta^{\bar\mu\nu} - \eta^{\mu\bar\mu}\eta^{\nu\bar\nu})}{\ell^2 - \lambda^2 + i\epsilon}$$

, and by replacing the QED charges by the corresponding gravity charges $\kappa k_{\bar{\mu}}$, $\kappa k'_{\bar{\nu}}$. This yields [13] for our scalar propagator

$$i\Delta'_F(k)|_{Resummed} = \frac{ie^{B''_g(k)}}{(k^2 - m^2 - \Sigma'_s + i\epsilon)}. \tag{7}$$

where

$$B''_g(k) = -2i\kappa^2 k^4 \frac{\int d^4\ell}{16\pi^4} \frac{1}{\ell^2 - \lambda^2 + i\epsilon} \frac{1}{(\ell^2 + 2\ell k + \Delta + i\epsilon)^2} \tag{8}$$

so that $B''_g(k) = \frac{\kappa^2|k^2|}{8\pi^2} \ln\left(\frac{m^2}{m^2+|k^2|}\right)$ in the deep Euclidean regime. If m vanishes, using the usual $-\mu^2$ normalization point we get $B''_g(k) = \frac{\kappa^2|k^2|}{8\pi^2} \ln\left(\frac{\mu^2}{|k^2|}\right)$. In both cases the resummed propagator falls faster than any power of $|k^2|$! This is the basic result. Note that Σ'_s starts in $\mathcal{O}(\kappa^2)$, so we may drop it in calculating the one-loop effects which are at the heart of our black hole physics analysis. Our result that the resummed propagator falls faster than any power of $|k^2|$ means that one-loop corrections are finite! Indeed, all quantum gravity loops are UV finite and the all orders proof, as well as the explicit finiteness of Σ'_s at one-loop, is given in Refs. [13].

4 Massive Elementary Particles and Black Hole Physics in RQG

The one-loop corrections to Newton's law implied by the diagrams in Fig. 1 are central to our discussion, as they directly impact our black hole physics issues. Using the YFS resummed propagators in Fig. 1, as we have shown in Refs. [13,21] we get the Newtonian potential $\Phi_N(r) = -\frac{G_N M_1 M_2}{r}(1-e^{-ar})$ where [13,21] $a \simeq 3.96 M_{Pl}$ when for definiteness we set $m \cong 120$GeV. Our gauge invariant analysis can be shown [13] to be consistent with the one-loop analysis of QG in Ref. [27][3].

With reasonable estimates and measurements [13, 28, 29] of the SM particle masses, including the various bosons, the corresponding results for the analogs of the diagrams in Fig. 1 imply [13] that in the SM $a_{eff} \cong 0.349 M_{Pl}$. To make direct contact with black hole physics, note that, if r_S is the Schwarzschild radius, for $r \rightarrow r_S$, $a_{eff} r \ll 1$ so that $|2\Phi_N(r)|_{m_1=m}/m_2| \ll 1$. This means that $g_{00} \cong 1 + 2\Phi_N(r)|_{m_1=m}/m_2$ remains positive as we pass through the Schwarzschild radius. It can be shown [13] that this positivity holds to $r = 0$. Similarly, g_{rr} remains negative through r_S down to $r = 0$ [13]. To get these results, note that in the relevant regime for r, the smallness of the quantum corrected newton potential means that we can use the linearized Einstein equations for a small spherically symmetric static source $\rho(r)$ which generates $\phi_{newton}(r)|_{m_1=m}/m_2$ via the standard Poisson's equation. the usual result (see Refs. [1, 30]) for the respective metric solution then gives $g_{00} \cong 1 + 2\phi_{newton}(r)|_{m_1=m}/m_2$ and $g_{rr} \cong -1 + 2\phi_{newton}(r)|_{m_1=m}/m_2$ which remain respectively time-like and space-like to $r = 0$.

In resummed QG, a massive SM point particle is not a black hole.

---

[3]Our deep Euclidean studies are complementary to the low energy studies of Ref. [22].

The value of a_{eff} given here is incomplete, as there may be as yet unknown massive particles beyond those already discovered – these would only decrease a_{eff}. For example, in the minimal supersymmetric Standard Model we expect approximately that $a_{eff} \to \frac{1}{\sqrt{2}}a_{eff}$.

One can also use the results for the complete one-loop UV divergent corrections of Ref. [27] to see that the remaining interactions at one-loop order not discussed here (vertex corrections, pure gravity self-energy corrections, etc.) also do not increase the value of a_{eff} – a_{eff} is a parameter which is bounded from above by the estimates we give here. In general, we expect that the precise value of a_{eff} should be determined from cosmological and/or other considerations. Such implications will be taken up elsewhere.

Our results imply the running Newton constant $G_N(k) = G_N/(1 + \frac{k^2}{a_{eff}^2})$ which is fixed point behavior for $k^2 \to \infty$, in agreement with the phenomenological asymptotic safety approach of Ref. [20]. Our result that an elementary particle has no horizon also agrees with the result in Ref. [20] that a black hole with a mass less than $M_{cr} \sim M_{Pl}$ has no horizon. The basic physics is the same: $G_N(k)$ vanishes for $k^2 \to \infty$.

Because our value of the coefficient of k^2 in the denominator of $G_N(k)$ agrees with that found by Ref. [20], if we use their prescription for the relationship between k and r in the regime where the lapse function vanishes, we get the same Hawking radiation phenomenology as they do: a very massive black hole evaporates until it reaches a mass $M_{cr} \sim M_{pl}$ at which the Bekenstein-Hawking temperature vanishes, leaving a Planck scale remnant.

We can carry this argument further as follows. In Ref. [20] there is some uncertainty about the precise value of M_{cr}; for, this value depends on a parameter γ which varies between 0 and $9/2$. However, independent of the value of γ, the result that the value of M_{cr} is of the order of M_{Pl} means that the quantum corrections to the Newtonian potential found above must be taken into account in the analysis of Ref. [20] when the mass of the black hole approaches this critical value in the Hawking process. When we do this, we find that the horizon is obviated [31]. This would mean that eventually the entire mass of the originally massive black hole would become accessible to the rest of the universe, which is consistent with Hawking's recent conclusions [32] on the very important subject of information loss in black hole physics.

5 Conclusions

YFS resummation renders quantum gravity finite so that quantum loop corrections are now cut off dynamically. We call the resultant new approach to quantum gravity resummed quantum gravity (RQG), which represents a minimal union of the ideas of Bohr and Einstein. Physics below the Planck scale is accessible to point particle quantum field theory. Early universe studies may be able to test some of our predictions [4] and we point-out that

[4] This is controversial [33] and is under investigation.

a theoretical cross check with the analysis of Ref. [27] has been done. In the discussion presented here, we focused on some consequences of RQG for black hole physics.

We have shown that, contrary to classical expectations, a massive elementary SM particle is not a black hole in resummed quantum gravity. Our results are also consistent with the asymptotic safety analysis in Ref. [20] that a black hole of mass less than a critical mass $\sim M_{Pl}$ does not have a horizon in quantum gravity and that the final state of the Hawking radiation of a massive black hole is a Planck scale remnant, which our recent work now shows becomes accessible to the rest of the universe due to our finite quantum loop corrections as well, a result which agrees with Hawking's recent results [32] on information loss in black hole physics. Further checks are under investigation.

Acknowledgements

We thank Profs. S. Bethke and L. Stodolsky for the support and kind hospitality of the MPI, Munich, while a part of this work was completed. We thank Prof. C. Prescott for the kind hospitality of SLAC Group A while this work was in progress. We thank Prof. S. Jadach for useful discussions.

References

[1] C. Misner, K.S. Thorne and J.A. Wheeler, *Gravitation*,(Freeman, San Francisco, 1973).

[2] S. Weinberg, *Gravitation and Cosmology: Principles and Applications of the General Theory of Relativity*,(John Wiley, New York, 1972).

[3] S.L. Glashow, *Nucl. Phys.* **22** (1961) 579; S. Weinberg, *Phys. Rev. Lett.* **19** (1967) 1264; A. Salam, in *Elementary Particle Theory*, ed. N. Svartholm (Almqvist and Wiksells, Stockholm, 1968), p. 367; G. 't Hooft and M. Veltman, *Nucl. Phys.* **B44**,189 (1972) and B50, 318 (1972); G. 't Hooft, *ibid.* B35, 167 (1971); M. Veltman, *ibid.* B7, 637 (1968).

[4] D. J. Gross and F. Wilczek, *Phys. Rev. Lett.* **30** (1973) 1343; H. David Politzer, *ibid.* 30 (1973) 1346; see also , for example, F. Wilczek, in *Proc. 16th International Symposium on Lepton and Photon Interactions, Ithaca, 1993*, eds. P. Drell and D.L. Rubin (AIP, NY, 1994) p. 593, and references therein.

[5] S. Hawking, *Nature (London)* **248** (1974) 30; *Commun. Math. Phys.* **43** (1975) 199.

[6] R. P. Feynman, *Acta Phys. Pol.* **24** (1963) 697.

[7] R. P. Feynman, *Feynman Lectures on Gravitation*, eds. F.B. Moringo and W.G. Wagner (Caltech, Pasadena, 1971).

[8] C. Itzykson and J.-B. Zuber, *Quantum Field Theory*, (McGraw-Hill, New York, 1980); M. Peskin and D.V. Schroeder, *An Introduction to Quantum Field Theory*, (Addison-Wesley,Menlo Park, 1995).

[9] S. Weinberg, *The Quantum Theory of Fields, vols. 1-3*,(Cambridge University Press, Cambridge, 1995,1996,2000).

[10] D. R. Yennie, S. C. Frautschi, and H. Suura, *Ann. Phys.* **13** (1961) 379; see also K. T. Mahanthappa, *Phys. Rev.* **126** (1962) 329, for a related analysis.

[11] See S. Jadach *et al.*,*Comput. Phys. Commun.* **102**,229 (1997); S. Jadach, M. Skrzypek and B.F.L. Ward,*Phys.Rev.D* **55**,1206 (1997); S. Jadach, B.F.L. Ward and Z. Was,*Phys. Rev.D* **63**, 113009 (2001); S. Jadach, B.F.L. Ward and Z. Was, *Comp. Phys. Commun.* **130**, 260 (2000); S. Jadach *et al., ibid.***140** 432, 475, (2001).

[12] B.F.L. Ward and S. Jadach, in *Proc. ICHEP02*, eds. S. Bentvelsen et al.,(North Holland, Amsterdam, 2003) p.275; *Acta. Phys. Pol.* **B33** (2002) 1543, and references therein.

[13] B.F.L. Ward, Mod. *Phys. Lett. A***17** (2002) 2371; *ibid.***19** (2004)143; *J. Cos. Astropart. Phys.***0402**, 011 (2004).

[14] See, for example, M. Green, J. Schwarz and E. Witten, *Superstring Theory, v. 1 and v.2*, (Cambridge Univ. Press, Cambridge, 1987) and references therein.

[15] See, for example, J. Polchinski, *String Theory, v. 1 and v. 2*, (Cambridge Univ. Press, Cambridge, 1998), and references therein.

[16] See for example V.N. Melnikov, *Gravit. Cosmol.* **9** (2003) 118; L. Smolin, hep-th/03-03185, and references therein.

[17] D.J. Gross, hep-th/9704139, in *Proc. Conceptual Foundations of Quantum Field Theory*, ed. T.Y. Cao (Cambridge Univ. Pr., Cambridge, 1999) p. 56, an references therein.

[18] S. Weinberg, in *General Relativity*, eds. S.W. Hawking and W. Israel,(Cambridge Univ. Press, Cambridge, 1979) p.790.

[19] O. Lauscher and M. Reuter, hep-th/0205062, and references therein.

[20] A. Bonanno and M. Reuter, *Phys. Rev. D***62** (2000) 043008.

[21] B.F.L. Ward, hep-ph/0410273, in *Proc. ICHEP2004*, in press, 2005.

[22] See for example J. Donoghue, *Phys. Rev. Lett.* **72** (1994) 2996; *Phys. Rev. D***50** (1994) 3874; preprint gr-qc/9512024; J. Donoghue *et al.*, *Phys. Lett. B* **529** (2002) 132, and references therein.

[23] D. Bardin and G. Passarino,*The Standard Model in the Making : Precision Study of the Electroweak Interactions* , (Oxford Univ. Press, London, 1999).

[24] D. Abbaneo *et al.*, hep-ex/0212036; see also, M. Gruenewald, hep-ex/0210003, in *Proc. ICHEP02*, in press, 2003.

[25] S. Perlmutter *et al.*, *Astrophys. J.* **517** (1999) 565; and, references therein.

[26] L. D. Faddeev and V.N. Popov, ITF-67-036, NAL-THY-57 (translated from Russian by D. Gordon and B.W. Lee); *Phys. Lett. B***25** (1967) 29.

[27] G. 't Hooft and M. Veltman, *Ann. Inst. Henri Poincare XX*, **69** (1974).

[28] See for example D. Wark, in *Proc. ICHEP02*, in press; and, M. C. Gonzalez-Garcia, hep-ph/0211054, in *Proc. ICHEP02*, in press, and references therein.

[29] K. Hagiwara *et al.*, *Phys. Rev. D***66** (2002) 010001; see also H. Leutwyler and J. Gasser, *Phys. Rept.* **87** (1982) 77, and references therein.

[30] R. Adler, M. Bazin and M. Schiffer, *Introduction to General Relativity* ,(McGraw-Hill, New York, 1965).

[31] B.F.L. Ward, to appear.

[32] S.W. Hawking, in *Proc. GR17*, 2004, in press.

[33] E. W. Kolb, talk presented at University of Munich, 2002.

In: Focus on Black Hole Research
Editor: Paul V. Kreitler, pp. 105-119
ISBN 1-59454-460-3

Chapter 3

QUASINORMAL MODES, REDUCED PHASE SPACE AND AREA SPECTRUM OF BLACK HOLES

M.R. Setare*
Physics Dept. Inst. for Studies in Theo. Physics and Mathematics(IPM),
P. O. Box 19395-5531, Tehran, IRAN

Abstract

Motivated by the recent interest in quantization of black hole area spectrum, we consider the area spectrum of Schwarzschild, BTZ, extremal Reissner-Nordström, near extremal Schwarzschild-de Sitter, and Kerr black holes. Based on the proposal by Bekenstein and others that the black hole area spectrum is discrete and equally spaced, we implement Kunstatter's method to derive the area spectrum for these black holes. We show that although as Schwarzschild black hole the spectrum is discrete, it is non equispaced in general. In the other hand the reduced phase space quantization is another technique which we discuss here. However there is a discrepancy between the result of the reduced phase space methodology and quasinormal modes approach for area spectrum of some black holes.

1 Introduction

Hawking was the first to observe that black holes are not completely black but they emit radiation [1]. This radiation is essentially thermal and the black hole emits field quanta of all frequencies. Now, may be one can ask this question: do quantum black holes have a discrete spectrum? Several authors have raised the possibility that Hawking radiation might in fact have a discrete spectrum. This idea was first proposed by Bekenstein in 1974 [2]. According to Bekenstein's proposal, the eigenvalues of the black hole event horizon area are of the form

$$A_n = \alpha l_p^2 n \ , \tag{1}$$

*E-mail address: rezakord@ipm.ir

where α is a dimensionless constant, n ranges over positive integers and, by using gravitational units $G = c = 1$, $l_p = \hbar^{1/2}$ is the Planck length. The black hole entropy-law ($S = \frac{A}{4\hbar}$) combined with a statistical interpretation of the entropy requires that $\exp A4\hbar$ is an integer. This, in turn, constraints the spacing constant parameter α in Eq.(1) to take the form $\alpha = 4\ln b$ for some integer b. Endorsement for Bekenstein's proposal was provided by the observation that the area of the horizon A behaves, for a slowly changing black hole, as an adiabatic invariant [3]. It is significant that a classical adiabatic invariant corresponds to a quantum observable with a discrete spectrum, by virtue of Ehrenfest's principle.
Since the original heuristic arguments of Bekenstein there has been a substantial amount of work in trying to derive the spectrum (1) by more rigorous means [2, 5]. Of particular relevance to the upcoming analysis is a program that was initiated by Barvinsky and Kunstatter [6]. Their methodology is based on expressing the black hole dynamics in terms of a reduced phase space and then applying an appropriate process of quantization. One vital assumption was required in this analysis; namely, the authors assumed that the conjugate to the mass is periodic over an interval of $\frac{2\pi}{k}$ where the k is the surface gravity at the horizon. They did, however, justify this input by way of Euclidean considerations.
In the other hand, recently, the quantization of the black hole area has been considered [7, 8] as a result of the absorption of a quasi-normal mode excitation[9, 10]. Any non-dissipative systems has modes of vibrations, which forming a complete set, and called normal modes. Each mode having a given real frequency of oscillation and being independent of any other. The system once disturbed continues to vibrate in one or several of the normal modes. On the other hand, when one deals with open dissipative system, as a black hole, instead of normal modes, one considers quasi-normal modes for which the frequencies are no longer pure real, showing that the system is loosing energy. The possibility of a connection between the quasinormal frequencies of black holes and the quantum properties of the entropy spectrum was first observed by Bekenstein [11], and further developed by Hod [7]. In particular, Hod proposed that the real part of the quasinormal frequencies, in the infinite damping limit, might be related via the correspondence principle to the fundamental quanta of mass and angular momentum.
In this paper we would like to discuss on the discrepancy between the reduced phase space methodology and QNM approach for area spectrum of some black holes.

2 Quasinormal Modes and Area Spectrum

It can be shown that for any periodic system, there exists an adiabatic invariant, which can be calculated (up to a constant shift) as follows

$$I \equiv \oint pdq \propto \int \frac{dE}{\omega(E)}, \tag{2}$$

where (p, q) are its phase space variables, E is the energy of system and $\omega(E)$ is vibrational frequency. Via Bohr-Sommerfeld quantization has an equally spaced spectrum in the semi-classical (large n) limit:

$$I \approx n\hbar. \tag{3}$$

From this viewpoint, the main problem of black hole quantum mechanics has been to correctly identify the physically relevant period of vibrational frequency. Recently Hod [7] assumed an equally spaced area spectrum and used the apparent existence of a unique quasinormal mode frequency in the large damping limit to uniquely fix the spacing.
In this section we use the observation about quasinormal mode frequency to drive the general form in the semi-classical limit of the Bekenstein-Hawking entropy spectrum for Schwarzschild, BTZ, extremal Reissner-Nordström, near extremal Schwarzschild-de Sitter, and Kerr black holes[17, 21, 22, 23].
We start from the observation that for a Schwarzschild black hole of mass M and radius R, the real part of the quasinormal mode frequency approaches a fixed non-zero value in the large damping limit as[15, 16]

$$\omega_R = \frac{\ln 3}{4\pi R} \tag{4}$$

Following [7, 8], Kunstatter in his recent interesting paper [17] assume that this classical frequency plays an important role in the dynamics of the black hole and is relevant to its quantum properties. He consider ω_R to be a fundamental vibrational frequency for a black hole of energy $E = M$. Using Eq.(2), we are lead to the following adiabatic invariant:

$$I = 4\pi \int \frac{dER}{\ln 3} = \frac{\hbar S_{BH}}{\ln 3} + c, \tag{5}$$

where we have used the fact that $R = 2M = 2E$ and the definition of the Bekenstein-Hawking entropy $S_{BH} = \frac{A}{4\hbar} = \frac{\pi R^2}{4\hbar}$. Eq.(3) then implies that the entropy spectrum is equally spaced:

$$S_{BH} = n \ln 3 \tag{6}$$

Then, the degeneracy of quantum states is given by:

$$\Omega(E) = \exp(S_{BH}) = 3^n \tag{7}$$

In this case also we obtain

$$\Delta A = 4\hbar \ln 3 \tag{8}$$

then the spacing constant parameter α in Eq.(1) fine as

$$\alpha = 4 \ln 3 \tag{9}$$

It has been suggested that the presence of $\ln 3$ implies a change in the loop quantum gravity gauge group from $SU(2)$ to $SO(3)$ [8]. Now we would like to obtain the area spectrum of non-rotating BTZ black hole. The non rotated BTZ black hole line element is as following [19, 20]

$$ds^2 = -(-M + \frac{r^2}{l^2})dt^2 + \frac{dr^2}{(-M + \frac{r^2}{l^2})} + r^2 d\theta^2, \tag{10}$$

which has an horizon at

$$r_+ = \sqrt{\frac{M}{\Lambda}}, \tag{11}$$

and is similar to Schwarzschild black hole with the important difference that it is not asymptotically flat but it has constant negative curvature. The quasi-normal frequencies for non-rotating BTZ black hole have been obtained by Cardoso and Lemos in [18]

$$\omega = \pm m - 2iM^{1/2}(n+1) \qquad n = 0, 1, 2, ... \tag{12}$$

where m is the angular quantum number. Let $\omega = \omega_R - i\omega_I$, now by taking ω_R as previous example, we have

$$I = \int \frac{dE}{\omega_R} = \int \frac{1}{\pm m} dM = \frac{M}{\pm m} + c, \tag{13}$$

where c is a constant, therefore the mass spectrum is equally spaced

$$\begin{aligned} M &= mn\hbar, \quad m \geq 0 \\ M &= -mn\hbar, \quad m < 0. \end{aligned} \tag{14}$$

Then for mass spacing we have

$$\Delta M = \pm m\hbar, \tag{15}$$

which is the fundamental quanta of black hole mass. As one can see the mass spectrum is equally spaced only for a fixed m. For different m there are multiplets with different values of spacing which is given by Eq.(28). On the other hand, the black hole horizon area is given by

$$A = 2\pi r_+. \tag{16}$$

Using Eq.(11) we obtain

$$A = 2\pi\sqrt{\frac{M}{\Lambda}}. \tag{17}$$

The Bohr-Sommerfeld quantization law and Eq.(40) then implies that the area spectrum is as following [21],

$$A_n = 2\pi\sqrt{\frac{nm\hbar}{\Lambda}}. \tag{18}$$

As one can see although the mass spectrum is equally spaced but the area spectrum is not equally spaced.

In [24] Birmingham et al have shown that the quantum mechanics of the rotating BTZ black hole is characterized by a Virasoro algebra at infinity. Identifying the real part of the quasi-normal frequencies with the fundamental quanta of black hole mass and angular momentum, they found that an elementary excitation corresponds exactly to a correctly quantized shift of the Virasoro generator L_0 or $\bar{L}_0$ in this algebra. Similar to [24]we have not found a quantization of horizon area. The result for the area of event horizon is not equally spaced, in contrast with area spectrum of Schwarzschild black hole.

Now we would like to obtain the area and entropy spectrum of extremal Reissner-Nordström (RN) black holes. The RN black hole's (event and inner) horizons are given in terms of the black hole parameters by

$$r_{\pm} = M \pm \sqrt{M^2 - Q^2}, \tag{19}$$

where M and Q are respectively mass and charge of black hole. In the extreme case these two horizons are coincides

$$r_{\pm} = M, \qquad M = Q. \tag{20}$$

According a very interesting conclusion follows [37](see also more recent paper [38]) , the real part of the quasinormal frequency for extremal RN black holes coincides with the Schwarzschild value

$$\omega_R^{RN} = \frac{\ln 3}{4\pi R_H}, \tag{21}$$

where

$$R_H = 2M. \tag{22}$$

Now by taking ω_R^{RN} in this context, we have

$$I = \int \frac{dE}{\omega_R^{RN}} = \int \frac{4\pi R_H}{\ln 3} dM = \frac{4\pi}{\ln 3} \int 2M dM = \frac{4\pi}{\ln 3} M^2 + c, \tag{23}$$

where c is a constant. In the other hand, the black hole horizon area is given by

$$A = 4\pi r_+^2. \tag{24}$$

Which using Eq.(20) in extremal case is as following

$$A = 4\pi M^2. \tag{25}$$

The Boher-Sommerfeld quantization law and Eq.(23) then implies that the area spectrum is equally spaced,

$$A_n = n\hbar \ln 3. \tag{26}$$

By another method we can obtain above result. From Eq.(25) we get

$$\Delta A = 8\pi M \Delta M = 8\pi M \hbar \omega_R^{RN} \tag{27}$$

where we have associated the energy spacing with a frequency through $\Delta M = \Delta E = \hbar\omega_R^{RN}$. Now using Eqs.(21, 22)we have

$$\Delta A = \hbar \ln 3, \tag{28}$$

therefore the extremal RN black hole have a discrete spectrum as

$$A_n = n\hbar \ln 3. \tag{29}$$

Which is exactly the result of Eq.(26). Using the definition of the Bekenstein-Hawking entropy we have

$$S = \frac{A_n}{4\hbar} = \frac{n \ln 3}{4}. \tag{30}$$

Now our aim is to obtain the area and entropy spectrum of near extremal Schwarzschild- de Sitter black holes. The metric of the Schwarzschild-de Sitter spacetime is as following

$$ds^2 = -f(r)dt^2 + f^{-1}(r)dr^2 + r^2 d\Omega^2, \tag{31}$$

where

$$f(r) = 1 - \frac{2M}{r} - \frac{r^2}{a^2}, \tag{32}$$

where M is the mass of black hole, and a denoting the de Sitter curvature radius, which related to the cosmological constant by $a^2 = \frac{3}{\Lambda}$. The spacetime possesses two horizons, the black hole horizon is at $r = r_b$, and the cosmological horizon is at $r = r_c$, with $r_b < r_c$. The third zero of $f(r)$ locates at $r_0 = -(r_b + r_c)$. It is useful to express M and a^2 as function of r_b, r_c

$$a^2 = r_b^2 + r_b r_c + r_c^2 \tag{33}$$

and

$$2Ma^2 = r_b r_c (r_b + r_c) \tag{34}$$

The surface gravity k_b associated with the black hole horizon, as defined by the relation $k_b = \frac{1}{2}\frac{df}{dr}|_{r=r_b}$. Easily one can show

$$k_b = \frac{(r_c - r_b)(r_b - r_0)}{2a^2 r_b}. \tag{35}$$

Let us now specialize to the near extremal SdS black hole, which is defined as the spacetime for which the cosmological horizon is very close to the black hole horizon. According a recent study [25](based on identifying the relevant scattering potential with that of the Poscl-Teller model [26, 27]) the quasinormal spectrum of SdS space depends strongly on the orbital angular momentum of the perturbation field. Moreover, in the near extremal SdS case, the real part of the frequency, goes almost linearly with orbital angular momentum. In a more recent paper [28], the quasinormal mode spectrum was calculated for SdS space by way of the monodromy method [29]. When this form of the spectrum is then subjected to the near extremal limit, as was done explicitly in [28], there is absolutely no orbital angular momentum dependence in evidence. In recent paper by Medved and Martin [30] a possible resolution for this discrepancy have provided. According to this paper, the monodromy-based calculation, although perfectly valid in the non-degenerate regime, can not necessarily be extrapolated up to the point of horizon coincidence.

The analytical quasinormal mode spectrum for the near extremal SdS black hole has been derived by Cardoso and Lemos [31] is as following

$$\omega = k_b[-(n + 1/2)i + \sqrt{\frac{v_0}{k_b^2} - 1/4 + o[\Delta]}], \qquad n = 0, 1, 2, ... \tag{36}$$

where

$$v_0 = k_b^2 l(l+1), \tag{37}$$

for scalar and electromagnetic perturbations, and

$$v_0 = k_b^2 (l+2)(l-1), \tag{38}$$

for gravitational perturbations, l is the angular quantum number. In Eq.(59) Δ is squeezing parameter [30] which is as following

$$\Delta = \frac{r_c - r_b}{r_b} \ll 1. \tag{39}$$

Now by taking ω_R in this context, we have

$$I = \int \frac{dE}{\omega_R} = \int \frac{dM}{k_b\sqrt{\frac{v_0}{k_b^2} - 1/4}} = \frac{M}{k_b\sqrt{\frac{v_0}{k_b^2} - 1/4}} + c, \tag{40}$$

where c is a constant. Boher-Sommerfeld quantization then implies that the mass spectrum is equally spaced,

$$M = n\hbar k_b\sqrt{\frac{v_0}{k_b^2} - 1/4}. \tag{41}$$

The use of Eqs.(34,52) leads us to

$$\delta M = \frac{r_b \Delta r \delta r_b}{2a^2} = \hbar k_b\sqrt{\frac{v_0}{k_b^2} - 1/4}. \tag{42}$$

In the other hand, the black hole horizon area is given by

$$A_b = 4\pi r_b^2, \tag{43}$$

by the variation of the black hole horizon and use of Eq.(42) we have

$$\delta A_b = 8\pi r_b \delta r_b = 8\pi \frac{2a^2\hbar k_b\sqrt{\frac{v_0}{k_b^2} - 1/4}}{\Delta r}. \tag{44}$$

Now by using Eq.(56), one can obtain

$$\delta A_b = 24\pi\hbar\sqrt{\frac{v_0}{k_b^2} - 1/4}. \tag{45}$$

Similar result for δA_b by another method have been obtained in[32]. Then we can obtain the quantization of a near extremal SdS black hole area as

$$A_b = 24\pi n\hbar\sqrt{\frac{v_0}{k_b^2} - 1/4}. \tag{46}$$

Using the definition of the Bekenstein-Hawking entropy we have

$$S = \frac{A_b}{4\hbar} = 6\pi n\sqrt{\frac{v_0}{k_b^2} - 1/4}. \tag{47}$$

Now we extend directly the Kunstatter's approach [17] to determine mass and area spectrum of Kerr and extreme Kerr black holes. The metric of a four-dimensional Kerr black hole given in Boyer-Lindquist coordinates is

$$ds^2 = -(1 - \frac{2Mr}{\Sigma})dt^2 - \frac{4Mar\sin^2\theta}{\Sigma}dtd\varphi + \frac{\Sigma}{\Delta}dr^2 + \Sigma d\theta^2 + (r^2 + a^2 + 2Ma^2 r\sin^2\theta)\sin^2\theta d\varphi^2 \tag{48}$$

where

$$\Delta = r^2 - 2Mr + a^2 \tag{49}$$

$$\Sigma = r^2 + a^2\cos\theta \tag{50}$$

and M is the mass of black hole. The roots of Δ are given by

$$r_\pm = M \pm \sqrt{M^2 - a^2} \tag{51}$$

where r_+ is the radius of the event (outer) black hole horizon and r_- is the radius of the inner black hole horizon. In addition, we have defined the specific angular momentum as

$$a = \frac{J}{M} \tag{52}$$

where J is the angular momentum of the black hole. The Kerr black hole is rotating with angular velocity

$$\Omega = \frac{a}{r_+^2 + a^2} \tag{53}$$

$$= \frac{J}{2M\left(M^2 + \sqrt{M^4 - J^2}\right)} \tag{54}$$

which has been evaluated on the event black hole horizon. In gravitational units, the Kerr black hole horizon area and its Hawking temperature are given, respectively, by

$$A = 4\pi(r_+^2 + a^2) \tag{55}$$

$$= 8\pi\left(M^2 + \sqrt{M^4 - J^2}\right) \tag{56}$$

and

$$T_H = \frac{r_+ - r_-}{A} \tag{57}$$

$$= \frac{\sqrt{M^4 - J^2}}{4\pi M\left(M^2 + \sqrt{M^4 - J^2}\right)}. \tag{58}$$

By applying Bohr's correspondence principle, Hod[33] conjectured that the real part of the asymptotic quasinormal frequencies of Kerr black hole is given by the formula

$$\omega_R = T_H \ln 3 + m\Omega, \tag{59}$$

where m is the azimuthal eigenvalue of the oscillation. There was compelling evidence that the conjectured formula (59) must be wrong.
A systematic exploration of the behavior of the first few overtones, i.e., small values of the principal quantum number n, was first accomplished by Onozawa [34]. Onozawa used the Leaver's continued fraction method to carry out the numerical calculations. Berti and Kokkotas [35] confirmed Onozawa's results and extended them to higher overtones, i.e., highly damped QNMs. They found that the formula conjectured by Hod, i.e., equation (59), does not seem to provide a good fit to the asymptotic modes. Furthermore, Berti and Kokkotas showed that, as the mode order increases, modes having their orbital angular momentum eigenvalue l and azimuthal eigenvalue m to satisfy $l = m = 2$, are fitted extremely well by the relation[1]

$$\omega = 2\Omega + i2\pi T_H n \ . \tag{60}$$

where the term $2\pi T_H$ that appears in the imaginary part of the mode frequencies is the surface gravity of the event horizon of the Kerr black hole. Therefore, the real part of the asymptotic frequencies having $l = m = 2$ is given by the expression

$$\omega_R = 2\Omega \ . \tag{61}$$

The first law of black hole thermodynamics now takes the form

$$dM = \frac{1}{4} T_H dA + \Omega dJ \tag{62}$$

and obviously the corresponding expression for adiabatic invariant is now given by the expression

$$I = \int \frac{dM - \Omega dJ}{\omega_R} \ . \tag{63}$$

The real part (ω_R) of the asymptotic (highly damped) quasinormal frequencies of the Kerr black hole is given by equation (71) and the angular velocity is given by equation (54). Therefore, the adiabatically invariant integral (63) is written as

$$I = \frac{2}{mJ} \int M \left(M^2 + \sqrt{M^4 - J^2} \right) dM - \frac{1}{m} \int dJ \tag{64}$$

[1] A more sophisticated study of highly damped Kerr QNMs is performed in [35]. In this work the authors provide complementing and clarifying results that were presented in previous works[34, 35].

and after integration, we get

$$I = \frac{1}{2mJ}\left[M^2\left(M^2 + \sqrt{M^4 - J^2}\right) - J^2 ln\left(M^2 + \sqrt{M^4 - J^2}\right)\right] - \frac{1}{m}J \ . \qquad (65)$$

By equating expressions (3) and (65), we get

$$M^2\bar{A} - J^2 ln\bar{A} - c = \left(2mJl_p^2\right)n \qquad (66)$$

where the parameter c is equal to $2J^2$, the quantity $\bar{A}$ is the reduced horizon area

$$\bar{A} = \frac{A}{8\pi} \qquad (67)$$

and the area A of the Kerr black hole horizon is given by equation (56). The solution to equation (66) is the principal branch of Lambert W-function and thus the area of the Kerr black hole is written

$$A = 8\pi\left(-\frac{J^2}{M^2}\right)W_0[z] \qquad (68)$$

where the argument of Lambert W-function is given by

$$z = \left(-\frac{M^2}{J^2}\right)e^{-2\left(1+\frac{m\hbar}{J}n\right)} \ . \qquad (69)$$

Since we are only interested in highly damped quasinormal frequencies, i.e. $n \to \infty$, we keep only the first term of the series expansion of the Lambert W-function and we get

$$A = 8\pi e^{-2\left(1+\frac{m\hbar}{J}n\right)} \qquad (70)$$

It is obvious that the area spectrum, although discrete, is not equivalently spaced even to first order.

Hod studied again analytically the QNMs of Kerr black hole [36] and he concluded that the asymptotic quasinormal frequencies of Kerr black hole are given by the simpler expression

$$\omega_R = m\Omega \qquad (71)$$

which is obviously in agreement with the aforesaid numerical results of Berti and Kokkotas. This classical frequency plays an important role in the dynamics of the black hole and is relevant to its quantum properties [7, 8].

3 Reduced Phase Space and Area Spectrum

We now consider a specific reduced phase space model of charged black hole [6] (see also [12]). The starting point of this model is the result of [13]that the dynamics of static spherically symmetric charged configuration in any classical theory of gravity in $d-$spacetime dimensions is governed by an effective action of the form

$$\Gamma = \int dt(P_M\dot{M}c^2 + P_Q\dot{Q} - H(M,Q)), \qquad (72)$$

where M and Q are the mass and the charge respectively and P_M, P_Q the corresponding conjugate momenta. The momentum P_M has the interpretation of asymptotic time difference between the left and right wedges of a Kruskal diagram. Note that H is independent of P_M and P_Q, such that from Hamilton's equations, M and Q are constant of motion. Now, to incorporate thermodynamics of black holes, one assumes that the conjugate momentum P_M is periodic with period equal to inverse Hawking temperature times $\hbar$. That is,

$$P_M \sim P_M + \frac{\hbar}{k_B T_H(M,Q)}. \tag{73}$$

Similar assumptions were made in the past using different arguments [14]. Note that the above identification implies that the (M, P_M) phase subspace has a wedge removed from it, which makes it difficult, if not possible to quantize on the full phase-space. Thus, one can make a canonical transformation $(M, Q, P_M, P_Q) \to (X, Q, \Pi_X, \Pi_Q)$, which on the one hand opens up the phase space, and on the other hand, naturally incorporate the periodicity Eq.(73)[13]

$$X = \sqrt{\frac{\hbar(S_{BH} - S_0(Q))}{\pi k_B}} \cos(2\pi P_M k_B T_H/\hbar) \tag{74}$$

$$\Pi_X = \sqrt{\frac{\hbar(S_{BH} - S_0(Q))}{\pi k_B}} \sin(2\pi P_M k_B T_H/\hbar) \tag{75}$$

$$Q = Q \tag{76}$$

$$\Pi_Q = P_Q + \phi P_M + S_0' P_M T_H/k_B \tag{77}$$

where the entropy at extremality is given by $S_0(Q) = \frac{\pi k_B Q^2}{\hbar c}$, $' \equiv \frac{d}{dQ}$ and ϕ is the electrostatic potential at the horizon. The new phase space is R^4, on which a rigorous quantization can be performed in a straightforward fashion. Moreover, as shown in [6, 12] it is straightforward to calculate the adiabatic invariant for charged black holes in this parameterization. In particular, Eqs.(47,26) generalize to the following invariant:

$$I \equiv \oint \Pi_X dX = \frac{\hbar(S_{BH} - S_0(Q))}{2\pi k_B} = (2n+1)\hbar \tag{78}$$

Quantization yields the following spectra for entropy and charge of the four dimensional quantum black hole(for more details refer to the [12]):

$$S_{BH} = (2n + p + 1)\pi k_B, \qquad n = 0, 1, 2, ... \tag{79}$$

$$\frac{Q}{e} = m, \qquad m = 0, \pm 1, \pm 2, ... \tag{80}$$

$$p \equiv \frac{Q^2}{\hbar c} = m^2 \frac{e^2}{\hbar c} \tag{81}$$

where consistency requires p to be a non-negative integer.
Recently the above formalism was extended to rotating (uncharged) black holes in 4 space-time dimensions [39]. The authors showed that the corresponding area spectrum is also discrete and equispaced:

$$A = 8\hbar\pi(n + m + 1/2) \tag{82}$$

where n and m are non-negative integers, while n signifies the departure of the black hole from extremality, m measures the classical angular momentum of the black hole. More recently, Das et al [40], extend their formalism to include BTZ black holes as well as 5 and higher dimensional Myeres-Perry type rotating black holes with multiple angular momenta parameters. They show that while the area spectrum is discrete in each case in general it is not equispaced.

4 Conclusions

Bekenestein's idea for quantizing a black hole is based on the fact that its horizon area, in the nonextremal case, behaves as a classical adiabatic invariant. Discrete spectra arise in quantum mechanics in the presence of a periodicity in the classical system Which in turn leads to the existence of an adiabatic invariant or action variable. Boher-Somerfeld quantization implies that this adiabatic invariant has an equally spaced spectrum in the semi-classical limit. In this article we have reviewed some recent study in black hole area spectrum, including Kunstatter's approach, and the reduced phase space methodology. We have extended the Kunstatter's approach for Schwarzschild black hole to another black holes including, BTZ, extremal Reissner-Nordström, near extremal Schwarzschild-de Sitter, and Kerr black holes, and have shown that although as Schwarzschild black hole the spectrum is discrete, it is non equispaced in general. We have shown that while the area spectrum is discrete in each case in general it is not equispaced. The result for the area of event horizon in BTZ black hole case is as: $A_n = 2\pi\sqrt{\frac{nm\hbar}{\Lambda}}$ which is not equally spaced. Then we have found that the quantum area for extremal RN black hole in four dimensional spacetime, is as: $\Delta A = \hbar \ln 3$. Although the real parts of highly damped quasi-normal modes for Schwarzschild and extremal RN black hole is equal [38] $\omega_R = \frac{\ln 3}{8\pi M}$, as one can see for example in [17, 41, 38], $\Delta A = 4\hbar \ln 3$ for Schwarzschild black hole. Therefore in contrast with claim of [38], ΔA is not universal for all black holes. Also we have found that the quantum area for near extremal SdS black hole in four dimensional spacetime is as: $\Delta A = 24\pi\hbar\sqrt{\frac{v_0}{k_b^2} - 1/4}$. Also Abdalla et al [32] have shown that the results for spacing of the area spectrum for near extreme Kerr black holes differ from that for Schwarzschild, as well as for non-extreme Kerr black holes. Such a difference for problem under consideration in this letter also in [32] as the authors have been mentioned may be justified due to the quite different nature of the asymptotic quasi-normal mode spectrum of the near extreme black hole. Finally we have shown that the area spectrum for Kerr black hole is as: $A = 8\pi e^{-2\left(1+\frac{m\hbar}{J}n\right)}$, it is obvious that the area spectrum, although discrete, is not equivalently spaced.
In the other hand the reduced phase space quantization is another technique which we have

discussed here. However there is a discrepancy between the result of the reduced phase space methodology and QNM approach for area spectrum of some black holes. As for why there is an apparent discrepancy between the QNM calculations and that of reduced phase space methodology, there is ne easy answer. It is not clear what role (if any) quasinormal spectrum plays in quantum theory of gravity (which as of now is nonexistent). I see only two useful applications of quasinormal spectrum: classical general relativity and AdS/CFT correspondence [42], where QNM appear as poles of the retarded correlations functions in the quantum field theory dual to a particular gravitational background [43]

References

[1] S. W. Hawking. *Commun. Math. Phys.* 1975, 43, 199-220.

[2] J. D. Bekenstein. *Lett. Nuovo Cimento.* 1974, 11, 467.

[3] J. D. Bekenstein. *Phys. Rev. D.* 1973, 7, 2333-2346.

[4] J. D. Bekenstein ; G. Gour. *Phys. Rev. D.* 2002. 66, 024005.

[5] G. Gour. *Phys. Rev. D.* 2002. 66, 104022.

[6] A. Barvinsky ; G. Kunstatter. (1996). *Mass Spectrum for Black Holes in Generic 2-D Dilaton Gravity* [gr-qc/9607030].

[7] S. Hod. *Phys. Rev. Lett.* 1998, 81, 4293.

[8] O. Dreyer. *Phys. Rev. Lett.* 2003, 90, 081301.

[9] T. Regge ; J. A. Wheeler. *Phys. Rev.* 1957, 108, 1063.

[10] S. Chandrasekhar ; S. Detweiler. *Proc. R. Soc. London, Ser. A.* 1975, 344, 441.

[11] J. D. Bekenstein. (1997). Quantum Black Holes as Atom [gr-qc/9710076].

[12] A. Barvinsky ; A. Das, G. Kunstatter. *Class. Quant. Grav.* 2001, 18,4845-4862. and *Phys. Lett. B.* 2001, 517, 415-420.

[13] D. Louis-Martinez ; G. Kunstatter, *Phys. Rev. D.* 1995, 52, 3494-3505.

[14] H. Kastrup, *Phys. Lett. B.* 1996, 385, 75-80. J. Louko ; J. Makela, *Phys. Rev. D.* 1996, 54, 4982-4996.

[15] K. D. Kokkotas ; B. G. Schmidt, *Living Rev. Rel.* 1999, 2, 2.

[16] H. P. Nollert, *Class. Quant. Grav.* 1999, 16, R159.

[17] G. Kunstatter, *Phys. Rev. Lett.* 2003, 90, 161301.

[18] V. Cardoso ; J. P. S. Lemos, *Phys. Rev. D.* 2001, 63, 124015.

[19] M. Bañados; C. Teitelboim ; J. Zanelli, *Phys. Rev. Lett.* 1992, 69, 1849-1851.

[20] M. Bañados, M. Henneaux, C. Teitelboim and J. Zanelli, *Phys. Rev. D.* 1993, 48, 1506-1525.

[21] M. R. Setare, *Class. Quant. Grav.* 2004, 21, 1453-1458.

[22] M. R. Setare, *Phys. Rev. D.* 2004, 69, 044016.

[23] M. R. Setare. (2004). Near Extremal Schwarzschild-de Sitter Black Hole Area Spectrum from Quasi-normal Modes [hep-th/0401063].

[24] D. Birmingham; S. Carlip; Y. Chen. *Class. Quant. Grav.* 2003, 20 L239-L244.

[25] V. Cardoso ; J. P. S. Lemos. *Phys. Rev. D.* 2003, 67, 084020.

[26] G. Poshl ; E. Teller. *Z. Phys.* 1933, 83, 143.

[27] V. Ferrari ; B. Mashhoon. *Phys. Rev. D.* 1984, 30, 295.

[28] K. H. C. Castello-Branco ; E. Abdalla. (2003). Analytic determination of the asymptotic quasi-normal mode spectrum of small Schwarzschild-de Sitter black holes [gr-qc/0309090].

[29] L. Motl. *Adv. Theor. Math. Phys.* 2003, 6, 1135-1162.

[30] A. J. M. Medved ; Damien Martin. (2003). A Note on Quasinormal Modes: A tale of two treatments [gr-qc/0311086].

[31] V. Cardoso; R. A. Konoplya ; J. P. S. Lemos, *Phys. Rev. D.* 2003, 68, 044024.

[32] E. Abdalla; K. H. C. Castello-Branco ; A. Lima-Santos. *Mod. Phys. Lett. A.* 2003, 18, 1435-1440.

[33] S. Hod. *Phys. Rev. D.* 2003, 67, 081501.

[34] H. Onozawa. *Phys. Rev. D.* 1997, 55, 3593-3602.

[35] E. Berti ; K.D. Kokkotas. *Phys. Rev. D.* 2003, 68, 044027.

[36] S. Hod. (2003). Asymptotic Quasinormal Mode Spectrum of Rotating Black Holes [gr-qc/0307060].

[37] A. Neitzke. (2003). Greybody factors at large imaginary frequencies [hep-th/0304080].

[38] N. Andersson; C. J. Howls. *Class. Quant. Grav.* 2004, 21, 1623-1642.

[39] G. Gour ; A. J. M. Medved, *Class. Quant. Grav.* 2003, 20, 2261-2274.

[40] S. Das; H. Mokhopadhyay ; P. Ramadevi. (2004). Spectrum of rotating black holes and its implications for Hawking radiation [hep-th/0407151].

[41] D. Birmingham. *Phys. Lett. B.* 2003, 569, 199-203.

[42] J.M. Maldacena. *Adv. Theor. Math. Phys.* 1998, 2, 231-252.

[43] D. T. Son; A. O. Starinets. *JHEP.* 2002, 0209, 042.

In: Focus on Black Hole Research
Editor: Paul V. Kreitler, pp. 121-158
ISBN 1-59454-460-3

Chapter 4

WHAT BLACK HOLES CAN TEACH US

Sabine Hossenfelder*
Department of Physics, University of Arizona,
1118 East 4th Street, Tucson, AZ 85721, USA

Abstract

Black holes merge together different field of physics. From General Relativity over thermodynamics and quantum field theory, they do now also reach into the regime of particle and collider physics. In the presence of additional compactified dimensions, it would be possible to produce tiny black holes at future colliders. We would be able to test Planck scale physics and the onset of quantum gravity. The understanding of black hole physics is a key knowledge to the phenomenology of these new effects beyond the Standard Model.

1 Introduction

Gravity is different. Physicists have successfully described three of the known particle interactions in the Standard Model of particle physics. The electromagnetic interaction, which governs the motion of electrically charged particles, the weak interaction, which is e.g. responsible for the beta decay, and the strong interaction, which bounds together the nuclei that we are made of. These three forces can be described as quantum field theories whose mathematical traps and physical details are thought to be well known; if not understood then they are at least familiar.

Only gravity refuses to fit in the common treatment. That gravity is somewhat special can be experienced in daily life. It is the only long range interaction and thus governs the dynamics of macroscopic objects, such as apples, planets or galaxies. Gravity is by far weaker than the other interactions. The strength of the electromagnetic force between two electrons is $\approx 10^{40}$ larger than the gravitational one! The only reason why we do observe

*E-mail address: sabine@physics.arizona.edu

it at all is that the gravitational attraction can not be neutralized. The charge of gravity, the mass, is always positive. But for all purposes of particle physics at small distances, the other interactions dominate and gravity can, and is, safely neglected.

This article is organized as follows. In the next section we will introduce the Planck scale and motivate its importance. Section three then reviews briefly some of the main features of black holes, including the evaporation characteristics. In Section four we will summarize the effective models with extra space time dimensions and examine some of the predicted observables. Section five then deals with black holes in these models with extra dimensions, the possibility of their production in the laboratory, their properties and observables. Section six investigates the formation of stable black hole remnants. In Section 7 we connect black hole physic to the issue of a minimal length scale. Section 8 is dedicated to the recurring question whether the earth produced black holes are dangerous. We conclude with a brief summary in Section 9.

Unless otherwise indicated, we assume natural units $\hbar = c = 1$ which leads to $m_{\rm p} = 1/l_{\rm p}$. The metric has signature $(-1, 1, 1, \ldots)$.

2 The Planck Scale

For the description of particle physics in the laboratory, interactions of the Standard Model dominate and gravity can be safely neglected. But things will change dramatically if we go to extreme regimes. Let us consider a test particle that is accelerated to a very high energy E. As we know from quantum mechanics, such a particle can, and must be, associated with a Compton wave length $\lambda = 1/E$. On the other hand, Einstein's Theory of General Relativity tells us that every kind of energy will cause space time to curve. This is described by Einstein's Field Equations:

$$R_{\mu\nu} - \frac{1}{2} g_{\mu\nu} R = -8\pi \frac{1}{m_{\rm p}^2} T_{\mu\nu} \quad . \tag{1}$$

The sources of gravity are everything which carries energy and momentum. These quantities are described in the energy momentum tensor $T_{\mu\nu}$ on the right hand side. The sources cause a curvature of spacetime which is described by the metric $g_{\mu\nu}$, and some of its second derivatives in the curvature tensor[1] R. This in turn will then modify the motion of the sources in the very space time, giving a self consistent time evolution of the system.

The equation of motions for a pointlike particle is the Geodesic equation:

$$\frac{{\rm d}^2 x^\mu}{{\rm d}\tau^2} + \Gamma^\mu_{\alpha\beta} \frac{{\rm d}x^\alpha}{{\rm d}\tau} \frac{{\rm d}x^\beta}{{\rm d}\tau} = 0 \quad , \tag{2}$$

where the Christoffel-symbols Γ are defined in Eq. (58) and τ is the eigen time.

[1] See Appendix A, Eq.(59), (60) and (61).

The metric of space time has mass dimension 0. Since the curvature is a second derivative, is has mass dimension 2. But the energy momentum tensor in three spacelike dimensions has dimension $1+3$. To get the dimensions right in Eq.(1), we thus have to introduce a mass scale $m_{\rm p}$ with a power of 2, the so called Planck scale.

The exterior metric of an arbitrary spherical symmetric mass distribution with mass M is described by the Schwarzschild solution [1]:

$$ds^2 = g_{\mu\nu}dx^\mu dx^\nu = -\gamma(r)dt^2 + \gamma(r)^{-1}dr^2 + r^2 d\Omega^2 \quad , \tag{3}$$

where

$$\gamma(r) = 1 - \frac{1}{m_{\rm p}^2}\frac{2M}{r} \quad , \tag{4}$$

and $d\Omega$ is the surface element of the 3-dimensional unit sphere.

To identify the constant $m_{\rm p}$, we take the limit to the familiar Newtonian potential by considering a slow motion in a static gravitational field. Here, slow means slow relative to the speed of light in which case the eigen time is almost identical to the coordinate time, $dt/d\tau \approx 1$, and for the spatial coordinates $i = 1,2,3$ it is $dx^i/d\tau \approx 0$ in Eq.(2). This yields the approximation

$$\frac{d^2x^i}{d\tau^2} \approx \frac{d^2x^i}{dt^2} \approx -\Gamma^i_{tt}\left(\frac{dt}{d\tau}\right)^2 \approx -\Gamma^i_{tt} = g^{ij}\Gamma_{jtt} \quad . \tag{5}$$

Using the definition of the Christoffel-symbols Eq.(58) and taking into account the time independence of the gravitational field, $\partial_t g_{\mu\nu} \approx 0$, we obtain

$$\frac{d^2x^i}{dt^2} \approx -\frac{1}{2}g_{tt,i} \quad . \tag{6}$$

Inserting Eq. (4) then yields

$$\frac{d^2x^i}{dt^2} \approx -\frac{M}{m_{\rm p}^2}\frac{1}{r^2} \quad , \tag{7}$$

and we can identify the quantity γ with the Newtonian Potential ϕ by

$$\gamma(r) = 1 - 2\phi(r) \quad . \tag{8}$$

So, it is $m_{\rm p}^2 = 1/G$, the inverse of the Newtonian constant, and the Planck mass turns out to be $\approx 10^{16}$ TeV. Compared to the typical mass scales of the strong and electroweak interactions, which are ≈ 100 GeV, this is a huge value and reflects the fact that gravity is much weaker. This large gap between the mass scales of the Standard Model and gravity is also called the hierarchy problem.

Let us now come back to our test particle with a very high energy. Looking at Eq.(4), we see that the energy E inside a space time volume with side length given by the Compton wave length λ, will cause a perturbation of the metric which is of order $E^2/m_{\rm p}^2$. Since flat space is described by $\gamma = 1$, we can therefore draw the very important conclusion that a particle with an energy close to the Planck mass will cause a non negligible perturbation of space time on length scales comparable to its own Compton wavelength. Such a particle then should be subject to the yet unknown theory of quantum gravity. Physics at the Planck scale thus represents a future challenge, located between particle physics and General Relativity.

3 Black Holes

Black holes are an immediate consequence of the Schwarzschild solution Eq.(3). As we see, the metric component $g_{tt} = \gamma(r)$ has a zero at the radius $R_H = 2M/m_{\rm p}^2$, the so called Schwarzschild radius. For usual stellar objects like our sun is, this radius lies well inside the mass distributions where the exterior solution does no longer apply [2]. However, if the mass is so densely packed that it has a radius smaller than R_H, we will be confronted with the consequences of g_{tt} having a zero.

The effect is described most effective by looking at the redshift of a photon. This photon is send out with frequency ν_0 by an observer at radius r_0 and is received at a radius $r_1 > r_0$ with frequency ν_1. To exclude effects due to Doppler shift, we assume both, sender and receiver, are in rest. Frequencies count oscillations per time, so their relation is inverse to the relation of the eigen time intervals at both locations

$$\frac{{\rm d}\tau_1}{{\rm d}\tau_0} = \frac{\nu_0}{\nu_1} \quad . \tag{9}$$

Since both observers are in rest, it is ${\rm d}r^2 = {\rm d}\theta^2 = {\rm d}\phi^2 = 0$ on their world lines. The relation between the eigen times, ${\rm d}\tau_{0/1}^2 = -{\rm d}s_{0/1}^2$, and the coordinate times is given by the metric in Eq.(3)

$${\rm d}\tau_0^2 = \gamma(r_0){\rm d}t^2 \quad , \quad {\rm d}\tau_1^2 = \gamma(r_1){\rm d}t^2 \quad . \tag{10}$$

By taking the square root and inserting the eigen time intervals into Eq.(9) we find that the photon will be received at the location r_1 with a frequency

$$\nu_1 = \nu_0 \frac{\sqrt{\gamma(r_1)}}{\sqrt{\gamma(r_0)}} \quad . \tag{11}$$

This redshift goes to infinity for $r_0 \to R_H$, $\gamma(r_0) \to 0$. In this case, when the sender is close to the Schwarzschild radius, the photon can not be sent out, no matter how high its initial energy is. Thus, no information will ever reach the observer at r_1. More generally speaking, it turns out that at R_H the space time has a trapped surface. What holds for the photon must also hold for slower moving objects. Nothing can ever escape from the region inside the Schwarzschild radius. The object is completely black: a black hole. The radius R_H is also called the event horizon.

[2]The Schwarzschild radius for the sun is ≈ 3 km, whereas its actual radius is ≈ 100.000 km.

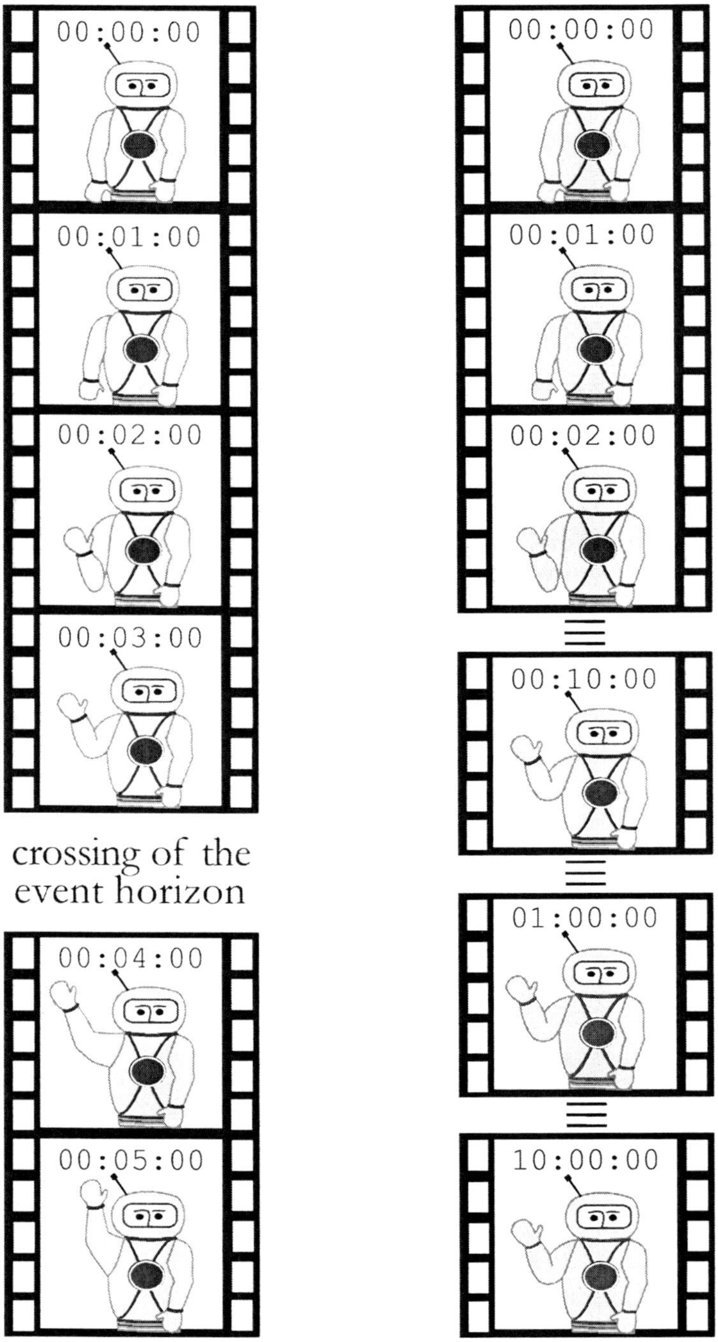

Figure 1: The last greetings from the astronaut who crosses the Schwarzschild radius. The left strip shows the pictures he sends out and the right strip shows the pictures that are received by an observer who is far away from the horizon.

It is noteworthy at this point that the existence of a horizon, weird at it might seem, is not troublesome in any regard. Even though the Schwarzschild metric in the above given form Eq.(3) has a pole in the g_{rr} component at R_H, this pole is not a physical one, meaning it can be removed by a suitable coordinate transformation. All of the curvature components are perfectly well behaved at the horizon.

The dependence of the eigen time intervals on the location of the observer can be illustrated by examining the information that is sent out by an infalling observer. Though the above derived expression is not appropriate in this case (the observer is not located at a fixed radius), the result is qualitatively the same, see e.g. [2]. The observer who approaches the horizon sends out information in time intervals which are of equal length in his eigen time as depicted in Figure 1, left. These pictures are received by an observer located at an arbitrary large radius. Here, space time is flat and the coordinate time is identical to the eigen time.

The pictures which are received by the observer far away from the horizon are shown in Figure 1, right. When looking at the last greetings from the infalling observer, the receiver will notice that the time intervals the infalling observer claimed to have equal length become longer and longer. Also, the wave lengths of the photons become longer and the colors are red shifted as derived above. But in addition, it seems to the receiver that every motion of his infalling friend becomes slow motion until it eventually freezes when the sender crosses the event horizon.

This of course does not only apply for infalling observers but also for the collapsing matter. For this reason, black holes have also been termed 'frozen stars'.

What can not be removed is the singularity at $r = 0$, which turns out to be a point with infinite curvature. Infinities in physics usually signal that we have missed some essential point in our mathematical treatment and, once again, indicate that the extreme regions of gravity are poorly understood.

3.1 Radiation of Black Holes

In 1975, Hawking published his computation about the radiation of black holes [3]. He showed that a black hole can emit particles if one takes into account the effects of quantum gravity in curved space time. Since the not so black hole of mass M is quasi-static (it is 'frozen'), the spectrum is a thermal spectrum and can be related to a temperature

$$T = \frac{\kappa}{2\pi} \quad , \quad \text{with} \quad \kappa = \frac{1}{2} \partial_r \gamma(r)\big|_{r=R_H} \quad , \tag{12}$$

where κ is the surface gravity of the black hole. If one inserts all constants one finds this temperature to be extremely small

$$T = \frac{1}{8\pi} \frac{c^2 m_{\rm p}^2}{k_B} \frac{1}{M} \approx 10^{-6} \frac{M_\odot}{M} [\mathrm{K}] \quad . \tag{13}$$

Here, $M_\odot \approx 10^{54}$ TeV is the mass of the sun.

The smaller the mass of the black hole, the higher is its temperature. It is remarkable that this formula for the first time connects all fundamental constants and joins gravitation, quantum field theory and thermodynamics.

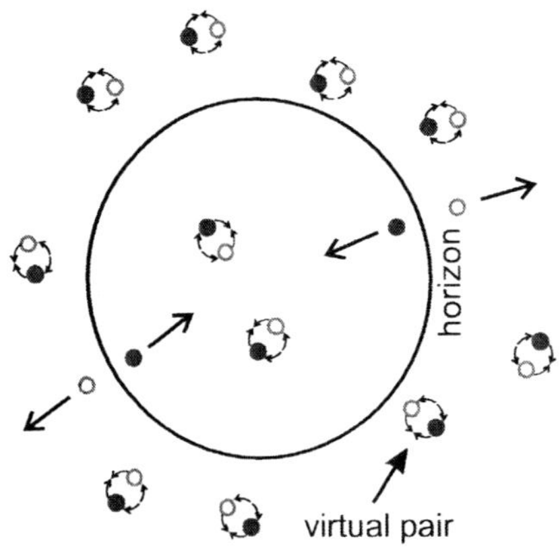

Figure 2: Particle creation at the horizon.

To understand this so called Hawking effect, we will use a simple analogy. The effect is similar to the particle production out of vacuum in quantum electrodynamics. In both cases, a strong external field is present. The vacuum is thought to be filled with virtual pairs of particles. In a background field strong enough, the virtual pairs can be ripped apart and form a real pair. The energy for this pair creation is provided by the background field. Figure 2 shows a schematic picture of this process. These analogies of course should not be put too far and can not replace a proper calculation. But they provide us with a phenomenological understanding of the effect.

One should be aware of two crucial differences to the electromagnetic case. One essential difference is that, in contrast to the electromagnetic case, in the gravitational case not the actual strength of the field is responsible for the pairs becoming real but its tidal forces are. This is due to the fact that the 'charges' of gravity are always positive. The second weak point of this analogy is that the energy of the gravitational field is not a well defined quantity. Since gravity is build on the fact that a local observer can not distinguish between gravity and acceleration, this energy is a coordinate dependent quantity and can e.g. be defined in the asymptotically flat coordinate system. In this system, the energy of the infalling particle is apparently negative (this is due to g_{tt} being negative, see Eq. (3)). The black hole therefore looses mass when one of the particles escapes and the other one falls in, as we would have expected.

This picture will be sufficient for our purposes. More details on the calculation can be found e.g. in [4, 5, 6].

The thermodynamic of black holes [2, 4] can be put even further and reveals amazing parallels by relating geometrical quantities of the hole to thermodynamic variables. The most important of this quantities is the entropy S.

We know from the standard thermodynamics that the integrating factor of the entropy is the inverse of the temperature. By identifying the total energy of the system with the mass of the black hole we find

$$\frac{\partial S}{\partial M} = \frac{1}{T} \quad . \tag{14}$$

After inserting Eq. (12)

$$\frac{\partial S}{\partial M} = 8\pi \frac{M}{m_{\rm p}^2} \tag{15}$$

and integrating we obtain

$$\begin{aligned} S(M) &= 4\pi \left(\frac{M}{m_{\rm p}}\right)^2 + \text{const.} \\ &= \frac{m_{\rm p}^2}{4}\mathcal{A} + \text{const.} \quad , \end{aligned} \tag{16}$$

where $\mathcal{A} = 4\pi R_H^2$ is the surface of the black hole. The additive constant is not relevant for our further treatment and we will set it to zero.

The black hole looses mass through the evaporation process. Eq. (13) then states that the temperature will be increased by this, resulting in an even faster mass loss. Using the above results, we can estimate the lifetime of the black hole under the assumption that it is a slow process. With all constants, the energy density of the radiation is then given by Stefan-Boltzmann's law

$$\varepsilon = \frac{\pi^2}{30\hbar^3 c^2}(k_B T_S)^4 \quad . \tag{17}$$

The variation of the mass of the black hole itself is then

$$\frac{\mathrm{d}M}{\mathrm{d}t} \approx -\varepsilon\mathcal{A} = -\frac{1}{15\pi 8^3}\frac{c^4\hbar}{G^2}\frac{1}{M_0^2} \quad , \tag{18}$$

where M_0 is the initial mass of the black hole under investigation. By integration over t we find for the lifetime

$$\tau \approx 15\pi 8^3 \frac{G^2}{c^4\hbar} M_0^3 \approx 10^{59}\left(\frac{M_0}{M_\odot}\right)^3 \text{Gyr} \quad , \tag{19}$$

after which the mass M is reduced to 0.

To get a grip on this quantity for astrophysical objects, let us insert some numbers. For black holes caused by collapse of usual astrophysical objects, it is $M_0 > 10^{15}$ g and τ exceeds the age of the universe ($\approx 10^{10}$ years). For super massive black holes, caused by secondary collapse of whole star clusters with typical masses of $M_0 \geq 10^{27}$ g, the temperature is even below the 3K cosmic background radiation and it does not evaporate anyhow. Mini black holes, which can be created by density fluctuations in the early universe would evaporate right now. The non-observation of this effects then sets limits on the spectrum of the primordial density fluctuations.

Looking back to our calculation we will notice that $\mathrm{d}M/\mathrm{d}t$ goes to infinity for late times. This unphysical result is due to the assumption that the black hole is treated as a heat bath and the back reaction which modifies the temperature by emission of a particle has been neglected. This estimation is pretty good for large and cold objects, such as astrophysical black holes. In general, however, this will no longer be the case if the mass of the black hole itself is close to its temperature. The late stages of evaporation will then be modified. We should not be surprised that the scale at which this will happen is when the mass of the black hole is close to the Planck mass, as can been seen from Eq. (13).

By applying the laws of statistical mechanics we can improve the calculation in such a way that it remains valid from the thermodynamical point of view even if the mass of the emitted particle gets close to the mass of the black hole. This can be done by using the micro canonical ensemble, as has been pointed out by Harms [7].

The Hawking result could also have been derived by using the canonical ensemble in which the number density of the particles (here, bosons) is given by

$$n(\omega) = \frac{1}{\exp\frac{\hbar\omega}{k_B T} - 1} \quad . \tag{20}$$

As familiar from the case of black body radiation this yields the total energy which we have used in (18) by integration over the 3 dimensional momentum space

$$\varepsilon = \frac{4\pi}{(2\pi)^3}\int_0^\infty n(\omega)\omega^3 \mathrm{d}\omega \propto T^4 \quad . \tag{21}$$

Here again, we use that the mass M is much larger than the energy of the emitted particles. Strictly spoken, it should not be possible to emit particles with $\omega > M$.

Now, instead of assuming the black hole being a heat bath, we use the micro canonical ensemble. Then, the number density for a single particle with energy ω and $\omega \leq M$ is

$$n(\omega) = \frac{\exp\left[S(M-\omega)\right]}{\exp[S(M)]} \quad , \tag{22}$$

where S is the entropy derived in (16).

The multi particle number density is

$$n(\omega) = \sum_{j=1}^{\lfloor \frac{M}{\omega} \rfloor} \frac{\exp[S(M-j\omega)]}{\exp[S(M)]} \quad , \tag{23}$$

where $\lfloor a \rfloor$ is the next smaller integer to a and assures that nothing can be emitted with energy above the mass of the black hole itself. This yields then the total number density

$$\varepsilon = \frac{4\pi}{(2\pi)^3}\mathrm{e}^{-S(M)}\int_0^\infty \sum_{j=1}^{\lfloor \frac{M}{\omega} \rfloor} \exp\left[S(M-j\omega)\right]\omega^3 \mathrm{d}\omega \quad . \tag{24}$$

We further substitute $x = (M - j\omega)$ (see also [8]). x is then the energy after emission of j particles of energy ω. After some algebra one finds

$$\varepsilon = \frac{4\pi}{(2\pi)^3} e^{-S(M)} \sum_{j=1}^{\infty} \frac{1}{j^4} \int_0^M e^{S(x)} (M-x)^3 dx \quad . \tag{25}$$

The sum has the value $\pi^4/90$. From this, the time dependence of the mass is given by

$$\frac{dM}{dt} = \frac{4\pi^3}{45} \frac{M^2}{m_p^4} \exp[-4\pi(M/m_p)^2] \int_0^M (M-x)^3 \exp[4\pi(x/m_p)^2] dx \quad . \tag{26}$$

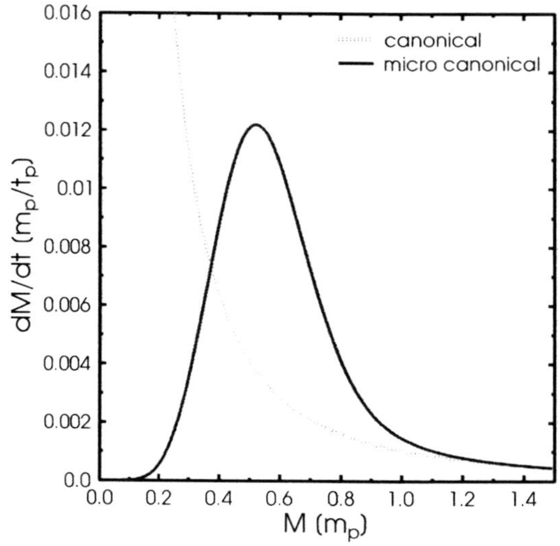

Figure 3: Canonical contra micro canonical.

Figure 3. shows this result in contrast to the canonical one for masses close to m_p. Here, t_p is the timescale associated with the Planck mass $t_p = \hbar/c^2 m_p$. As can be seen, for large M, both cases agree. For small M, the canonical evaporation rate diverges whereas the micro canonical evaporation rate drops for $M < m_p$ after passing a maximum close by m_p.

By integrating over dM/dt, we can again compute the lifetime. As a numerical example we investigate an initial mass of $M = 2m_p$. Here, the micro canonical approach yields $\tau \approx 1.45 \times 10^{17}\, m_p^{-1} \approx 10^{-3}fm/c$, whereas the canonical approach would yield $\tau \approx 2 \times 10^6\, m_p^{-1} \approx 10^{-14}fm/c$.

One might raise the objection that at the mass scales under consideration here, the effects of quantum gravity should play the most important role. Nevertheless, thermodynamics for quantum systems is a very powerful tool and unless we know more about quantum gravity we hope that it remains valid even close by the Planck scale.

In this section we have only considered the case in which the black hole is completely determined by the mass trapped in it. It turns out that even the most general vacuum solution of Einsteins Field Equations depends on only three parameters. Besides the mass M of the black hole, it can also carry angular momentum J and electric charge Q. The metric for this system is known as the Kerr-Newman metric [9], and the fact of the black hole solution only having a three parameter set is known as the no-hair theorem.

The result that every black hole is described by this three parameters means that during the collapse all other characteristics of the initial state, such as gravitational monopole moments, have to be radiated away. After this balding phase, the hole remains with its last three hairs: M, J and Q.

The angular momentum and the charge do also modify the evaporation spectrum which can be included in a straight forward way (see, e.g. [5, 10]). In case the initial state has angular momentum, the emitted radiation will carry away most of its rotational energy in a spin down phase.

4 Extra Dimensions

So far it might seem that Planck scale physics, though interesting, is only important at such high energies that it is out of reach for experimental examination on earth. This huge value of the Planck scale, however, might only be an apparent scale, caused by the presence of additional space like dimensions [11]. In contrast to our usual three space like dimensions, these additional dimensions are compactified to small radii which explains why we have not yet noticed them.

Motivated by String Theory, the proposed models with extra dimensions lower the Planck scale to values soon accessible. These models predict a vast number of quantum gravity effects at the lowered Planck scale, among them the production of TeV-mass black holes and gravitons.

4.1 The Model of Large Extra Dimensions

During the last decade, several models using compactified Large Extra Dimensions (LXDs) as an additional assumption to the quantum field theories of the Standard Model (SM) have been proposed. The setup of these effective models is motivated by String Theory though the question whether our spacetime has additional dimensions is well-founded on its own and worth the effort of examination.

The models with LXDs provide us with an useful description to predict first effects beyond the SM. They do by no means claim to be a theory of first principles or a candidate for a grand unification. Instead, their simplified framework allows the derivation of testable results which can in turn help us to gain insights about the underlying theory.

There are different ways to build a model of extra dimensional space-time. Here, we want to mention only the most common ones:

1. The ADD-model proposed by Arkani-Hamed, Dimopoulos and Dvali [12] adds d extra spacelike dimensions without curvature, in general each of them compactified to the same radius R. All SM particles are confined to our brane, while gravitons are allowed to propagate freely in the bulk.

2. The setting of the model from Randall and Sundrum [13, 14] is a 5-dimensional spacetime with an non-factorizable geometry. The solution for the metric is found by analyzing the solution of Einsteins field equations with an energy density on our brane, where the SM particles live. In the type I model [13] the extra dimension is compactified, in the type II model [14] it is infinite.

3. Within the model of universal extra dimensions [15] all particles (or in some extensions, only bosons) can propagate in the whole multi-dimensional spacetime. The extra dimensions are compactified on an orbifold to reproduce SM gauge degrees of freedom.

In the following we will focus on the model 1. which yields a beautiful and simple explanation of the hierarchy problem. Consider a particle of mass M located in a space time with $d+3$ dimensions. The general solution of Poisson's equation yields its potential as a function of the radial distance r to the source

$$\phi(r) \propto \frac{1}{M_{\rm f}^{d+2}} \frac{M}{r^{d+1}} \quad , \tag{27}$$

where we have introduced a new fundamental mass-scale[3] $M_{\rm f}$. The hierarchy problem then is the question why, for $d=0$, this mass-scale is the Planck mass, $m_{\rm p}$, and by a factor 10^{16} smaller than the mass-scales in the SM, e.g. the weak scale.

The additional d spacetime dimensions are compactified on radii R, which are small enough to have been unobserved so far. Then, at distances $r \gg R$, the extra dimensions will 'freeze out' and the potential Eq. (27) will turn into the common $1/r$ potential, but with a fore-factor given by the volume of the extra dimensions

$$\phi(r) \rightarrow \frac{1}{M_{\rm f}^{d+2}} \frac{1}{R^d} \frac{M}{r} \quad . \tag{28}$$

This is also sketched in Figure 4. In the limit of large distances, we will rediscover the usual gravitational law which yields the relation

$$m_{\rm p}^2 = M_{\rm f}^{d+2} R^d \quad . \tag{29}$$

Given that $M_{\rm f}$ has the right order of magnitude to be compatible with the other observed scales, it can be seen from this argument that the volume of the extra dimensions suppresses the fundamental scale and thus, explains the huge value of the Planck mass.

[3] Similar to our argument in Section 1, we note that the curvature still has dimension 2 but the energy momentum tensor now has dimension $1+3+d$. To get the dimensions right, we then have to introduce a mass scale with a power of $d+2$.

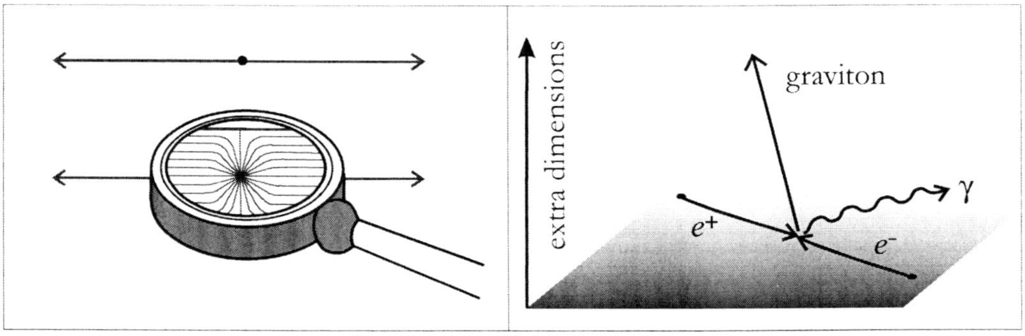

Figure 4: Left: at small distances the gravitational potential is higher dimensional. At large distances, we rediscover the three dimensional case. Right: schematic figure for a scattering process producing a graviton that escapes from our 3-dimensional submanifold and results in an energy loss.

The radius R of these extra dimensions, for $M_{\rm f} \sim$ TeV, can be estimated with Eq.(29) and typically lies in the range from 10^{-1} mm to 10^3 fm for d from 2 to 7, or the inverse radius $1/R$ lies in energy range eV to MeV, respectively. The case $d = 1$ is excluded. It would result in an extra dimension about the size of the solar system.

Due to the compactification, momenta in the direction of the LXDs can only occur in quantized steps $\propto 1/R$ for every particle which is allowed to enter the bulk. The fields can be expanded in Fourier-series

$$\psi(x,y) = \sum_{n=-\infty}^{+\infty} \psi^{(n)}(x) \exp\left(iny/R\right) \quad , \tag{30}$$

where x are the coordinates on our brane and y the coordinates of the LXDs. This yields an infinite number of equally spaced excitations, the so called Kaluza-Klein-Tower. On our brane, these massless KK-excitations act like massive particles, since the momentum in the extra dimensions generates an apparent mass term

$$\left[\partial_x \partial^x - \left(\frac{n}{R}\right)^2\right] \psi^{(n)}(x) = 0 \quad . \tag{31}$$

4.2 Observables of Extra Dimensions

The most obvious experimental test for the existence of extra dimensions is a measurement of the Newtonian potential at sub-mm distances. Cavendish like experiments which search for deviations from the $1/r$ potential have been performed during the last years with high precision [16] and require the extra dimensions to have radii not larger than $\sim 100\mu$m, which disfavors the case of $d = 2$.

Also the consequences for high energy experiments are intriguing. Since the masses of the KK-modes are so low, they get excited easily but it is not until energies of order $M_{\rm f}$ that their phase-space makes them give an important contribution in scattering processes. The number of excitations $N(\sqrt{s})$ below an energy $\sqrt{s}$ can, for an almost continuous spectrum, be estimated by integration over the volume of the d- dimensional sphere whose radius is the maximal possible wave number $n_{\rm max} = R\sqrt{s}$

$$N(\sqrt{s}) = \int_0^{n_{\rm max}} {\rm d}^d n = \Omega_{(d)}(R\sqrt{s})^d \quad . \tag{32}$$

We can then estimate the total cross-section for a point interaction, e.g. $e^+e^- \to$ graviton$+\gamma$ (depicted in Figure 6, right) by

$$\sigma(e^+e^- \to G\gamma) \approx \frac{\alpha}{m_{\rm p}^2} N(\sqrt{s}) \approx \frac{\alpha}{s}\left(\frac{\sqrt{s}}{M_{\rm f}}\right)^{d+2} \quad , \tag{33}$$

where we have used Eq.(29) and α is the fine structure constant. As can be seen, at energy scales close to the new fundamental scale, the estimated cross-section becomes comparable to cross-sections of electroweak processes.

The necessary Feynman rules for exact calculations of the graviton tree-level interactions have be derived [17] and the cross-sections have been examined closely. Since the gravitons are not detected, their emission would lead to an energy loss in the collision and to a higher number of monojets. Modifications of SM predictions do also arise by virtual graviton exchange, which gives additional contributions in the calculation of cross-sections.

Another exciting signature of LXDs is the possibility of black hole production as will be discussed in detail the next section. It is a direct consequence of the gravitational force being modified at distances smaller than the radius of the extra dimension.

These fascinating processes could be tested at the Large Hadron Collider (LHC), which is currently under construction at CERN and is scheduled to launch in September 2007. The LHC will collide proton beams with a c.o.m.[4] energy of $\sqrt{s} \approx 14$ TeV.

The here described effects would not only be observable in high energy collisions in the laboratory but also in cosmic ray events. Ultra high energetic cosmic rays (UHECRs) have been measured with energies up to $\approx 10^8$ TeV! Though this energy seems enormous, one has to keep in mind that the infalling UHECR hits an atmospheric nuclei in rest, which corresponds to a fixed target experiment. To compare to the energies reachable at colliders, the energy has to be converted in the c.o.m. system. A particle with a typical energy of $E \sim 10^7$ TeV in the laboratory frame which interacts with a nuclei of mass $m \sim 1$ GeV in rest corresponds to a c.o.m. energy $\approx \sqrt{2Em} \approx 100$ TeV. Nevertheless, this energy is still high enough to make UHECRs an important window to study first effects beyond the Standard Model. Possible observables have been examined e.g. in Refs. [18, 19].

Further, the emission of gravitons and black hole production leads to astrophysical consequences, such as enhanced cooling of supernovae and modification of the cosmic background radiation. The presently available data from collider physics as well as from astrophysics sets constraints on the parameters of the model. For a recent update see e.g. [20].

[4] center of mass

As we have seen, the LXD-model predicts a rich phenomenology and is an useful extension of the Standard Model that can be used to verify or falsify the existence of extra dimensions.

5 Black Holes in Extra Dimensions

In the standard $3+1$ dimensional space-time, the production of black holes requires a concentration of energy-density which can not be reached in the laboratory. As we have seen, in the higher dimensional space-time, gravity becomes stronger at small distances and therefore the event horizon is located at a larger radius.

However, for astrophysical objects we expect to find back the usual 3-dimensional Schwarzschild solution. In this case, the horizon radius is much larger than the radius of the extra dimensions and the influence of the extra dimensions is negligible. We, in the contrary, will be interested in the case where the black hole has a mass close to the new fundamental scale. This corresponds to a radius close to the inverse new fundamental scale, and thus $R_H \ll R$. Those two case are depicted in Figure 5.

In the case $R_H \ll R$ under investigation, the topology of the object can be assumed to be spherical symmetric $d+3$ dimensions. The boundary conditions from the compactification can be neglected. The hole is so small that it effectively does not notice the periodicity.

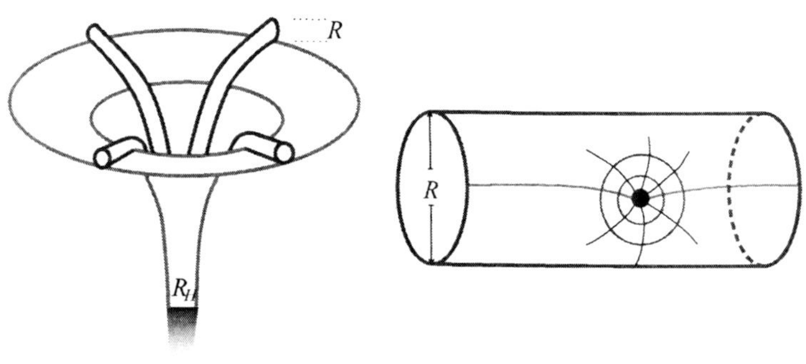

Figure 5: Sketch of the black hole topology in a space time with compactified extra dimensions. The left picture shows the astrophysical black hole with $R_H \gg R$, the right picture shows the collider produced black hole with $R_H \ll R$.

The higher dimensional Schwarzschild-metric has been derived in [21] and takes the form

$$ds^2 = -\gamma(r)dt^2 + \gamma^{-1}(r)dr^2 + r^2 d\Omega^2_{(d+3)} \quad , \tag{34}$$

where $d\Omega_{(d+3)}$ now is the surface element of the $3+d$ - dimensional unit sphere and

$$\gamma(r) = 1 - \left(\frac{R_H}{r}\right)^{d+1} \quad . \tag{35}$$

The constant R_H again can be found by requiring that we reproduce the Newtonian limit for $r \gg R$, that is $1/2\partial_r\gamma$ has to yield the Newtonian potential:

$$\frac{d+1}{2}\left(\frac{R_H}{r}\right)^{1+d}\frac{1}{r}=\frac{1}{M_{\rm f}^{d+2}}\frac{M}{r^{d+2}} \quad . \tag{36}$$

So we have

$$\gamma(r)=1-\frac{2}{d+1}\frac{1}{M_{\rm f}^{d+2}}\frac{M}{r^{d+1}} \quad , \tag{37}$$

and R_H is[5]

$$R_H^{d+1}=\frac{2}{d+1}\left(\frac{1}{M_{\rm f}}\right)^{d+1}\frac{M}{M_{\rm f}} \quad . \tag{38}$$

It is not surprising to see that a black hole with a mass about the new fundamental mass $M \sim M_{\rm f}$, has a radius of about the new fundamental length scale $L_{\rm f}=1/M_{\rm f}$ (which justifies the use of the limit $R_H \ll R$). For $M_{\rm f} \sim 1$TeV this radius is $\sim 10^{-4}$ fm. Thus, at the LHC it would be possible to bring particles closer together than their horizon. A black hole could be created.

The surface gravity can be computed using Eq.(12)

$$\kappa=\frac{1+d}{2}\frac{1}{R_H} \quad , \tag{39}$$

and the surface of the horizon is now given by

$$\mathcal{A} = \Omega_{(d+3)}R_H^{d+2} \tag{40}$$

$$= \Omega_{(d+3)}\frac{M}{\kappa}\left(\frac{1}{M_{\rm f}}\right)^{d+2} \quad , \tag{41}$$

where $\Omega_{(d+3)}$ is the surface of the $d+3$-dimensional unit sphere.

$$\Omega_{(d+3)}=\frac{2\pi^{\frac{d+3}{2}}}{\Gamma(\frac{d+3}{2})} \quad . \tag{42}$$

[5]Note, that this does not agree with the result found in [21] because in this case the extra dimensions were not compactified and therefore the constants were not matched to the four-dimensional ones. The fore-factors cancel in our case with the fore-factors from the higher dimensional Newtonian law. Though this is not in agreement with the relation used in most of the literature, it is actually the consequence of relation Eq. (29). These fore-factors, however, are of order one and can be absorbed in an redefinition of the new fundamental scale $M_{\rm f}$.

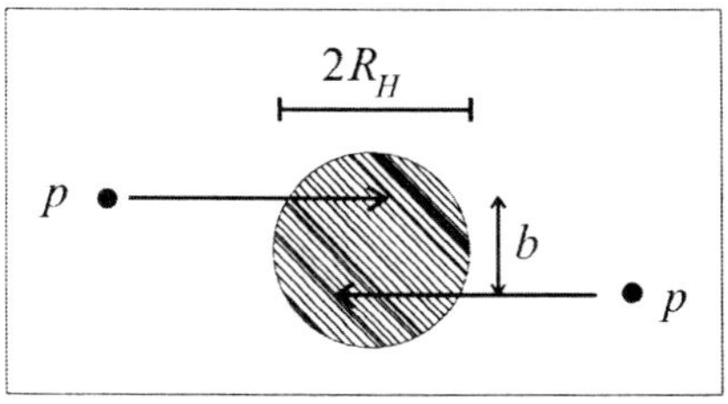

Figure 6: Schematic figure for a particle collision. At very high energies, the particles, p, can come closer than the Schwarzschild radius, R_H, associated with their energy. If the impact parameter, b, is sufficiently small, such a collision will inevitably generate a trapped surface. A black hole has been created.

5.1 Production of Black Holes

Let us consider two elementary particles, approaching each other with a very high kinetic energy in the c.o.m. system close to the new fundamental scale $M_{\mathrm{f}} \sim 1$ TeV, as depicted in Figure 6. At those high energies, the particles can come very close to each other since their high energy allows a tightly packed wave package despite the uncertainty relation. If the impact parameter is small enough, which will happen to a certain fraction of the particles, we have the two particles plus their large kinetic energy in a very small region of space time. If the region is smaller than the Schwarzschild radius connected with the energy of the partons, the system will collapse and form a black hole.

The production of a black hole in a high energy collision is probably the most inelastic process one might think of. Since the black hole is not an ordinary particle of the Standard Model and its correct quantum theoretical treatment is unknown, it is treated as a metastable state, which is produced and decays according to the semi classical formalism of black hole physics.

To compute the production details, the cross-section of the black holes can be approximated by the classical geometric cross-section

$$\sigma(M) \approx \pi R_H^2 \quad , \tag{43}$$

an expression which does only contain the fundamental Planck scale as coupling constant. This cross section has been under debate [22], but further investigations justify the use of the classical limit at least up to energies of $\approx 10M_{\mathrm{f}}$ [23]. However, the topic is still under discussion, see also the very recent contributions [24].

A common approach to improve the naive picture of colliding point particles, is to treat the creation of the horizon as a collision of two shock fronts in an Aichelburg-Sexl geometry describing the fast moving particles [25]. Due to the high velocity of the moving particles, space time before and after the shocks is almost flat and the geometry can be examined for the occurrence of trapped surfaces.

These semi classical considerations do also give rise to form factors which take into account that not the whole initial energy is captured behind the horizon. These factors have been calculated in [26], depend on the number of extra dimensions, and are of order one. For our further qualitative discussion we will neglect them. A time evolution of the system forming the black hole is not provided is this treatment [6] but it has been shown that the naively expected classical result remains valid also in String Theory [27].

Looking at Figure 6, we also see that, due to conservation laws, the angular momentum J of the formed object only vanishes in completely central collisions with zero impact parameter, $b = 0$. In the general case, we will have an angular momentum $J \approx 1/2Mb$. However, this modification turns out to be only about a factor 2 [28]. In general, the black hole will also carry a charge which gives rise to the exciting possibility of naked singularities. We will not further consider this case here, the interested reader is referred to [29].

Another assumption which goes into the production details is the existence of a threshold for the black hole formation. From General Relativistic arguments, two point like particles in a head on collision with zero impact parameter will *always* form a black hole, no matter how large or small their energy. At small energies, however, we expect this to be impossible due to the smearing of the wave functions by the uncertainty relation. This then results in a necessary minimal energy to allow for the required close approach. This threshold is of order $M_{\rm f}$, though the exact value is unknown since quantum gravity effects should play an important role for the wave functions of the colliding particles. For simplicity, we will in the following set this threshold equal to $M_{\rm f}$ but we want to point out that most of the analysis in the literature assumes a parameter of order one.

This threshold is also important as it defines the window in which other effects of the LXD scenario can be observed. Besides the graviton production and the excitation of KK-modes, also string excitations [30] have been examined. At even higher energies, highly exited long strings, the stringballs [31], are expected to dominate the scattering process. At the threshold energy, the so called correspondence point, the excited states can be seen either as strings or as black holes [32]. The smooth transition from strings to black holes has recently been investigated in Ref. [33]. The occurrence of these string excitations can increase the transverse size of the colliding objects and thus, make black hole production more difficult. This important point will be further discussed in Section 7.

Setting $M_{\rm f} \sim 1\,{\rm TeV}$ and $d = 2$ one finds $\sigma \sim 400$ pb. Using the geometrical cross section formula, it is now possible to compute the differential cross section $d\sigma/dM$ which will tell us how many black holes will be formed with a certain mass M at a c.o.m. energy $\sqrt{s}$. The probability that two colliding particles will form a black hole of mass M in a proton-proton collision at the LHC involves the parton distribution functions. These functions, $f_A(x, \hat{s})$, parametrize the probability of finding a constituent A of the proton (quark or gluon) with a momentum fraction x of the total energy of the proton. These constituents are also called partons. Here, $\sqrt{\hat{s}}$ is the c.o.m. energy of the parton-parton collision.

The differential cross section is then given by summation over all possible parton interactions and integration over the momentum fractions, where the kinematic relation $x_1 x_2 s =$

[6] If we think about it, this is in general also not provided in Standard Model processes.

$\hat{s} = M^2$ has to be fulfilled. This yields

$$\frac{\mathrm{d}\sigma}{\mathrm{d}M} = \sum_{A_1,B_2} \int_0^1 \mathrm{d}x_1 \frac{2\sqrt{\hat{s}}}{x_1 s} f_A(x_1,\hat{s}) f_B(x_2,\hat{s}) \sigma(M,d) \quad . \tag{44}$$

The particle distribution functions for f_A and f_B are tabulated e.g. in the CTEQ - tables. A numerical evaluation of this expression results in the differential cross section displayed in Figure 7, left. Most of the black holes have masses close to the production threshold. This is due to the fact that at high collision energies, or small distances respectively, the proton contents a high number of gluons which dominate the scattering process and distribute the total energy among themselves.

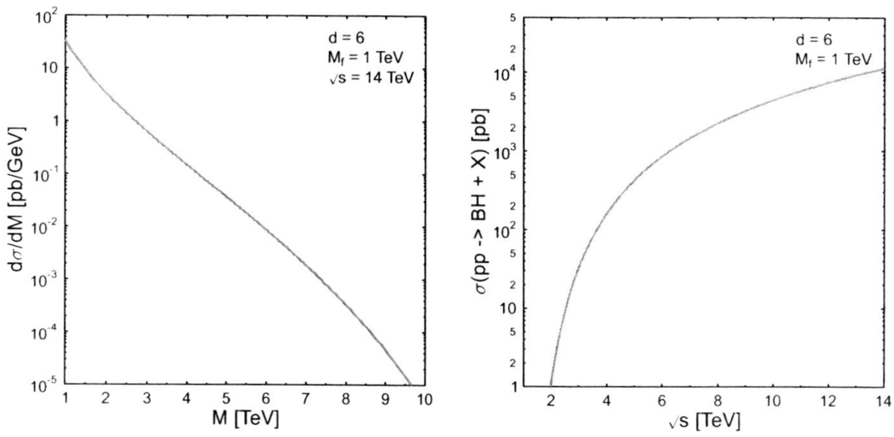

Figure 7: The left plot shows the differential cross section for black hole production in proton-proton-collisions at the LHC for $M_{\mathrm{f}} = 1$ TeV. The right plot shows the integrated total cross section as a function of the collision energy $\sqrt{s}$. In both cases, the curves for various d differ from the above depicted ones by less than a factor 10.

It is now straightforward to compute the total cross section by integration over Eq. (44), see Figure 7, right. This also allows us to estimate the total number of black holes, N_{BH}, that would be created at the LHC per year. It is $N_{\mathrm{BH}}/\mathrm{year} = \sigma(pp \to \mathrm{BH})L$ where we insert the estimated luminosity for the LHC, $L = 10^{33}\mathrm{cm}^{-2}\mathrm{s}^{-1}$. This yields at a c.o.m. of $\sqrt{s} = 14$ TeV a number of $N_{\mathrm{BH}} \approx 10^9$ per year! This means, about one black hole per second would be created. This result does also agree with the values found in [34]. The importance of this process led to a high number of publications on the topic of TeV-mass black holes at colliders [34, 35, 36, 37, 38, 39, 40, 41, 42].

The here presented calculation applies for the Large Hadron Collider and uses the parton distribution functions. At such high energies, these functions are extrapolations from low energy measurements and one might question whether the extrapolation can be applied into the regime of quantum gravity, or whether additional effects might modify them. Modifications might e.g. arise through the occurrence of a minimal length scale [43] or the presence of virtual graviton or KK-excitations. One should also keep in mind that the used extrapolations, the DGLAP[7] equations [44], rely on sliding scale arguments similar to the running of the coupling constants which is known to be modified in the presence of extra dimensions [45].

In this context it is therefore useful to examine black hole production at lepton colliders, which excludes the parton distribution functions as a source of uncertainty. A recent analysis has been given by in [42] for a muon collider. In general, the number for the expected black hole production at lepton colliders will be higher than for hadron colliders at equal energies, which is due to the collision energy not being distributed among the hadron constituents.

High energetic collision with energies close by or even above the new fundamental scale could also be reached in cosmic ray events [19]. Horizontally infalling neutrinos, which are scattered at nuclei in the atmosphere are expected to give the most important and clearest signature.

5.2 Evaporation of Black Holes

One of the primary observables in high energetic particle collisions is the transverse momentum of the outgoing particles, p_T, the component of the momentum transverse to the direction of the beam. Two colliding partons with high energy can produce a pair of outgoing particles, moving in opposite directions with high p_T but carrying a color charge, as depicted in Figure 8, top. Due to the quark confinement, the color has to be neutralized. This results in a shower of several bound states, the hadrons, which includes mesons (consisting of a quark and an antiquark, like the π's) as well as baryons (consisting of three quarks, like the neutron or the proton). The number of these produced hadrons and their energy depends on the energy of the initial partons. This process will cause a detector signal with a large number of hadrons inside a small opening angle. Such an event is called a 'jet'.

Typically these jets come in pairs of opposite direction. A smaller number of them can also be observed with three or more outgoing showers. This observable will be strongly influenced by the production of black holes.

To understand the signatures that are caused by the black holes we have to examine their evaporation properties. As we have seen before, the smaller the black hole, the larger is its temperature and so, the radiation of the discussed tiny black holes is the dominant signature caused by their presence. As we see from Eq. (39) and Eq. (13) the typical temperature lies in the range of several 100 GeV. Since most of the particles in the black body radiation are emitted with this average energy, we can estimate the total number of emitted particles to be of order 10-100. As we will see, this high temperature results in a very short lifetime such that the black hole will decay close by the collision region and can be interpreted as a

[7] Dokshitzer-Gribov-Lipatov-Altarelli-Parisi

metastable intermediate state.

The evaporation rate $\mathrm{d}M/\mathrm{d}t$ of the higher dimensional black hole can be computed using the thermodynamics of black holes. With Eq. (39), (41) and (12) we find for the temperature of the black hole

$$T = \frac{1+d}{4\pi}\frac{1}{R_H} \quad , \tag{45}$$

where R_H is a function of M by Eq. (38). Integration of the inverse temperature over M then yields (compare to Eq.(16))

$$S(M) \quad = \quad 2\pi\frac{d+1}{d+2}\left(M_{\mathrm{f}}R_H\right)^{d+2} \quad . \tag{46}$$

For the spectral energy density we use the micro canonical particle spectra Eq. (22) and (23) but we now have to integrate over the higher dimensional momentum space

$$\varepsilon = \frac{\Omega_{(d+3)}}{(2\pi)^{3+d}}\mathrm{e}^{-S(M)}\sum_{j=1}^{\infty}\frac{1}{j^{d+4}}\int_0^M \mathrm{e}^{S(x)}(M-x)^{3+d}\mathrm{d}x \quad , \tag{47}$$

where the value of the sum is given by a ζ-function. From this we obtain the evaporation rate

$$\frac{\mathrm{d}M}{\mathrm{d}t} = \frac{\Omega^2_{(d+3)}}{(2\pi)^{d+3}}R_H^{2+d}\zeta(4+d)\,\mathrm{e}^{-S(M)}\int_0^M (M-x)^{(3+d)}\mathrm{e}^{S(x)}\mathrm{d}x \quad . \tag{48}$$

A plot of this quantity for various d is shown in Figure 9, left. The mass dependence resulting from it is shown in Figure 9, right. We see that the evaporation process slows down in the late stages and enhances the life time of the black hole.

To perform a realistic simulation of the evaporation process, one has to take into account the various particles of the Standard Model with the corresponding degrees of freedom and spin statistics. In the extra dimensional scenario, Standard Model particles are bound to our submanifold whereas the gravitons are allowed to enter all dimensions. It has been argued that black holes emit mainly on the brane [35]. The ratio between the evaporation in the bulk and in the brane can be estimated by applying Eq. (48) to 3 or $3+d$ dimensions, respectively. For simplicity we will assume that the mass is high enough such that the result approximately agrees with the higher dimensional Stefan-Bolzmann law. Then it is

$$\frac{\mathrm{d}M}{\mathrm{d}t} \quad = \quad \frac{\Omega^2_{(3)}}{(2\pi)^3}R_H^2\zeta(4)\Gamma(4)\,T^4 \quad \text{on the brane, and} \tag{49}$$

$$\frac{\mathrm{d}M}{\mathrm{d}t} \quad = \quad \frac{\Omega^2_{(3+d)}}{(2\pi)^{3+d}}R_H^{2+d}\zeta(4+d)\Gamma(4+d)\,T^{4+d} \quad \text{in the bulk,} \tag{50}$$

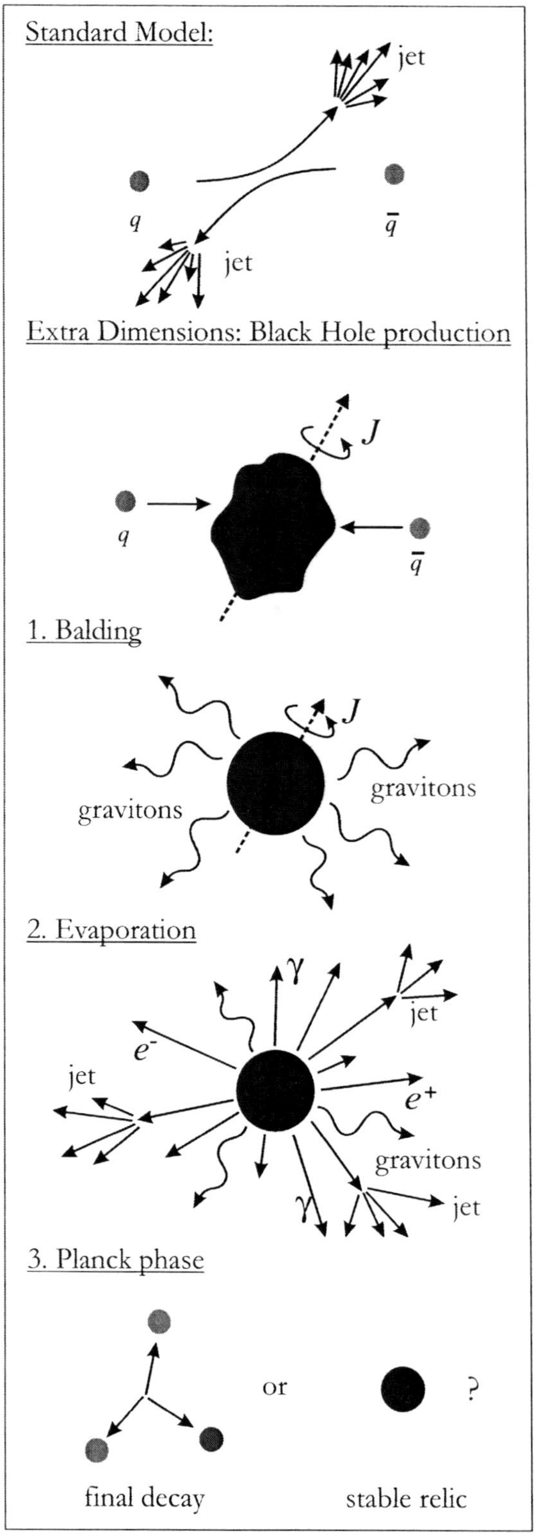

Figure 8: Phases of black hole evaporation.

Once produced, the black holes will undergo an evaporation process whose thermal properties carry information about the parameters M_{f} and d. An analysis of the evaporation will therefore offer the possibility to extract knowledge about the topology of our space time and the underlying theory.

The evaporation process can be categorized in three characteristic stages [36], see also the illustration in Figure 8:

1. BALDING PHASE: In this phase the black hole radiates away the multipole moments it has inherited from the initial configuration, and settles down in a hairless state. During this stage, a certain fraction of the initial mass will be lost in gravitational radiation.

2. EVAPORATION PHASE: The evaporation phase starts with a spin down phase in which the Hawking radiation carries away the angular momentum, after which it proceeds with emission of thermally distributed quanta until the black hole reaches Planck mass. The radiation spectrum contains all Standard Model particles, which are emitted on our brane, as well as gravitons, which are also emitted into the extra dimensions. It is expected that most of the initial energy is emitted in during this phase in Standard Model particles.

3. PLANCK PHASE: Once the black hole has reached a mass close to the Planck mass, it falls into the regime of quantum gravity and predictions become increasingly difficult. It is generally assumed that the black hole will either completely decay in some last few Standard Model particles or a stable remnant will be left, which carries away the remaining energy.

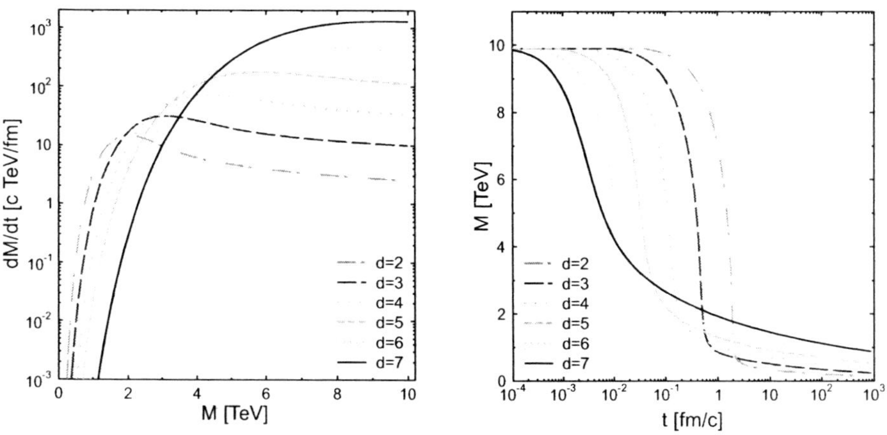

Figure 9: The left plot shows the evaporation rate in as a function of the initial mass for various d. The right plot shows the time evolution of a black hole with an initial mass of 10 TeV.

which yields a ratio brane/bulk of

$$\frac{\Omega^2_{(3)}}{\Omega^2_{(3+d)}} \frac{\zeta(4)}{\zeta(4+d)} \frac{\Gamma(4)}{\Gamma(4+d)} \left(\frac{2\pi}{TR_H}\right)^d \quad . \tag{51}$$

Inserting Eq. (45) results in a ratio, dependent on d, which is of order $4-12$, for $d \leq 7$ with a maximum[8] for $d=4$. In addition one has to take into account that there exists a much larger number of particles in the Standard Model than those which are allowed in the bulk. Counting up all leptons, quarks, gauge bosons, their anti particles and summing over various quantum numbers yields more than 60. This has to be contrasted with only one graviton in the bulk[9] which strongly indicates that most of the radiation goes into the brane.

However, the exact value will depend on the correct spin statistics and in general, the ratio also depends on the temperature of the black hole as can e.g. be seen in Figure 9, left. We also see that for masses close to the new fundamental scale the evaporation rate is higher for a smaller number of extra dimensions. This is due to the fact that in this region the evaporation rate is better approximated by a higher dimensional version of Wien's law. For a precise calculation one also has to take into account that the presence of the gravitational field will modify the radiation properties for higher angular momenta through backscattering at the potential well. These energy dependent greybody factors can be calculated by analyzing the wave equation in the higher dimensional spacetime and the arising

[8]This is due to the surface of the unit sphere having a maximum for 7 dimensions.

[9]In the effective description on the brane, one might argue that there exists a large number of gravitons with different masses. We have already taken this into account by integrating over the higher dimensional momentum space.

absorption coefficients. A very thorough description of these evaporation characteristics has been given in [10] which confirms the expectation that the bulk/brane evaporation rate is of comparable magnitude but the brane modes dominate.

For recent reviews on TeV-scale black holes see also [46] and references therein.

5.3 Observables of Black Holes

Due to the high energy captured in the black hole, the decay of such an object is a very spectacular event with a distinct signature. The number of decay products, the multiplicity, is high compared to Standard Model processes and the thermal properties of the black hole will yield a high sphericity of the event. Furthermore, crossing the threshold for black hole production causes a sharp cut-off for high energetic jets as those jets now end up as black holes instead, and are re-distributed into thermal particles of lower energies. Thus, black holes will give a clear signal.

It is apparent that the consequences of black hole production are quite disastrous for the future of collider physics. Once the collision energy crosses the threshold for black hole production, no further information about the structure of matter at small scales can be extracted. As it was put by Giddings and Thomas [36], this would be "*the end of short distance physics*".

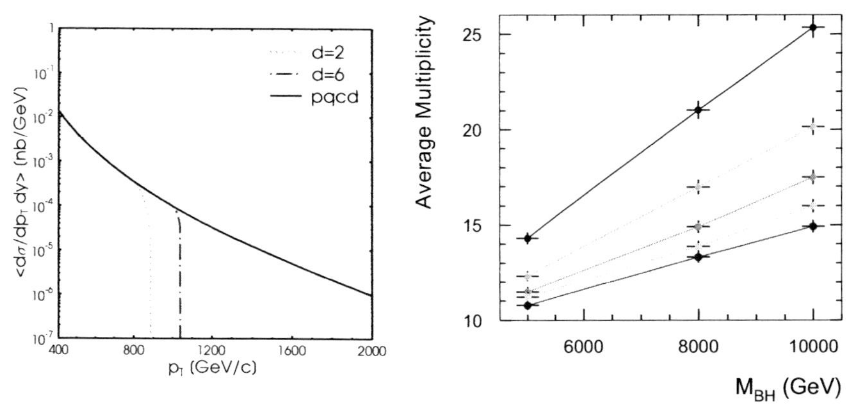

Figure 10: Left: The onset of black hole production in high energy collisions results in a cut off for high p_T-jets. Here, it is $\sqrt{s} = 14$ TeV and $M_f = 1$ TeV. The average over the rapidity region $|y| \leq 1$ is taken. Right: This plot from [41] shows the multiplicity of black hole events for various numbers of extra dimensions $d = 2$ (black), 3 (green), 4 (red), 5 (cyan) and 6 (blue).

Figure 10, left shows the expected cross section for jets at high transverse momentum with an onset of black hole production in comparison to the perturbative QCD prediction. The cutoff is close by the new fundamental scale [10] but is hardly sensitive to the number of

[10] We want to remind the reader that for qualitative arguments we have set the black hole threshold to be

extra dimensions.

By now, several experimental groups include black holes into their search for physics beyond the Standard Model. PYTHIA 6.2 [47] and the CHARYBDIS [48] event generator allow for a simulation of black hole events and data reconstruction from the decay products. Such analysis has been summarized in Ref. [40] and Ref. [41], respectively. Ideally, the energy distribution of the decay products allows a determination of the temperature (by fitting the energy spectrum to the predicted shape) as well as of the total mass of the object (by summing up all energies). This then allows to reconstruct the scale $M_{\rm f}$ and the number of extra dimensions. An example for this is shown in Figure 11, from [34].

The quality of the determination, however, depends on the uncertainties in the theoretical prediction as well as on the experimental limits e.g. background from Standard Model processes. Besides the formfactors of black hole production and the greybody factors of the evaporation, the largest theoretical uncertainties turn out to be the final decay and the time variation of the temperature. In case the black hole decays very fast, it can be questioned [34] whether it has time to readjust its temperature at all or whether it essentially decays completely with its initial temperature.

Also, the determination of the properties depends on the number of emitted particles. The less particles, the more difficult the analysis.

Figure 10 right, from [41], shows the number of expected decay products of a black hole event. For larger d with a fixed mass, the temperature increases which leads to less higher energetic particles. In Ref. [41] it was also pointed out that the high sphericity even at low multiplicity allows clear distinction from Standard Model and/or SUSY events. In a simulated test case, it was found that the fundamental scale can be determined up to 15% and the number of extra dimensions up to ± 0.75.

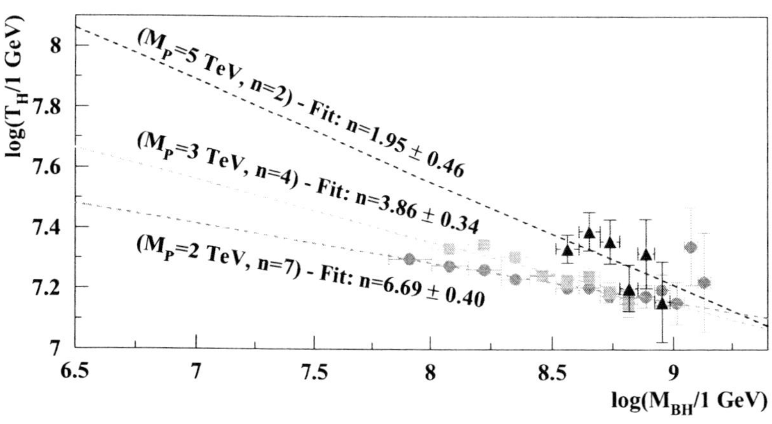

Figure 11: This plot from [34] shows an example for a reconstruction of the parameters from simulated black hole events. Here, it was assumed that the black hole decays so fast that the temperature can not be readjusted and is therefore time independent.

---

equal to the fundamental scale.

6 Black Hole Relics

The final fate of these small black holes is difficult to estimate. The last stages of the evaporation are closely connected to the information loss puzzle. The black hole emits thermal radiation, whose sole property is the temperature, regardless of the initial state of the collapsing matter. So, if the black hole completely decays into statistically distributed particles, unitarity can be violated. This happens when the initial state is a pure quantum state and then evolves into a mixed state [5, 49, 51].

When one tries to avoid the information loss problem two possibilities are left. The information is regained by some unknown mechanism or a stable black hole remnant is formed which keeps the information. Besides the fact that it is unclear in which way the information should escape the horizon [52] there are several other arguments for black hole relics [53]:

- The uncertainty relation: The Schwarzschild radius of a black hole with Planck mass is of the order of the Planck length. Since the Planck length is the wavelength corresponding to a particle of Planck mass, a problem arises when the mass of the black hole drops below Planck mass. Then one has trapped a higher mass, $M \geq M_{\rm f}$, inside a volume which is smaller than allowed by the uncertainty principle [54]. To avoid this problem, Zel'dovich has proposed that black holes with masses below Planck mass should be associated with stable elementary particles [55].

- Corrections to the Lagrangian: The introduction of additional terms, which are quadratic in the curvature, yields a dropping of the evaporation temperature towards zero [56, 57]. This holds also for extra dimensional scenarios [58] and is supported by calculations in the low energy limit of string theory [59, 60].

- Further reasons for the existence of relics have been suggested to be black holes with axionic charge [61], the modification of the Hawking temperature due to quantum hair [62] or magnetic monopoles [63]. Coupling of a dilaton field to gravity also yields relics, with detailed features depending on the dimension of space-time [64, 65].

Of course these relics, which have also been termed Maximons, Friedmons, Cornucopions, Planckons or Informons[11], are not a miraculous remedy but bring some problems on their own. Such is e.g. the necessity for an infinite number of states which allows the unbounded information content inherited from the initial state.

However, let us consider a very simple model which will help us to understand the features which lead to the formation of a stable black hole remnant. We will suppose that the spectrum of the radiation emitted by the black hole is discretized by geometrical means. The smaller the black hole is, the smaller is the region of space time out of which the radiation emerges and we expect the finite size of the black hole to reflect in the features of the radiation. The details of this ansatz have been worked out in [66].

[11] Though there are subtle but important differences between these objects.

In particular, for the derivation of the black hole's Hawking radiation, it turns out that the black hole acts like a black body. These analogy to geometrical optics can be put even further, see e.g. [50].

We split the solution for the $d+3$ dimensional wave equation of the black hole modes into an amplitude proportional to $\propto 1/r^{d+1}$, an angular dependence and a radial part. We then assume that the spectrum of the modes can only have discrete values due to the geometric boundary conditions $k_l = \pi l/R_H$, as depicted in Figure 12.

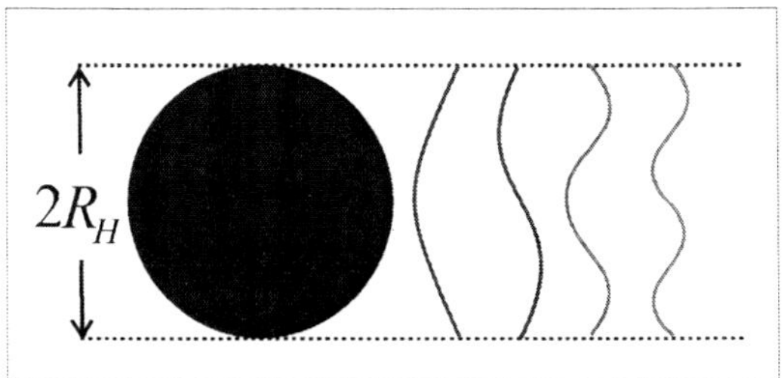

Figure 12: The black hole can not emit wave length that do not fit to the horizon size. The spectrum of the radiation is geometrically quantized in discrete steps. If the black hole shrinks, the lowest energetic mode can already exceed the total energy and thus, can not be emitted.

The particle spectrum Eq. (23) is then replaced by a spectrum for the discrete energy values $\omega_l = l\Delta\omega$ with step size $\Delta\omega = \pi/R_H$

$$n(l) = \sum_{j=1}^{\lfloor \frac{M}{l\Delta\omega} \rfloor} \frac{\exp[S(M - jl\Delta\omega)]}{\exp[S(M)]} \Theta(M - l\Delta\omega) \quad . \tag{52}$$

Here, the Heaviside-function Θ cuts off the sum when the energy of one particle ($j = 1$) exceeds the whole mass of the black hole. The energy density is then given by a summation instead of an integration

$$\varepsilon = \frac{\Omega_{(d+3)}}{(2\pi)^{3+d}} \Delta\omega \sum_{l=1}^{\lfloor \frac{M}{\Delta\omega} \rfloor} n(l)\,(l\Delta\omega)^{d+3} \quad , \tag{53}$$

which does as before result in an expression for the evaporation rate. For large masses, $M \gg M_{\rm f}$, the sum becomes an integral and Eq.(53) reduces to the continuous case. Figure 13, left, shows the evaporation rate with geometrical quantization in contrast to the continuous result (compare to Figure 9).

With increasing M, the steps between the energy levels become smaller. If it is possible to occupy an additional level, the evaporation rate has a step. In the mass region of interest for us, it is $R_H \approx 1/M_{\rm f}$ and thus the steps are $\approx \pi$TeV.

The lowest energetic mode possible corresponds to the largest wavelength which fits into the black holes horizon. The smaller the black hole is, the smaller this wavelength

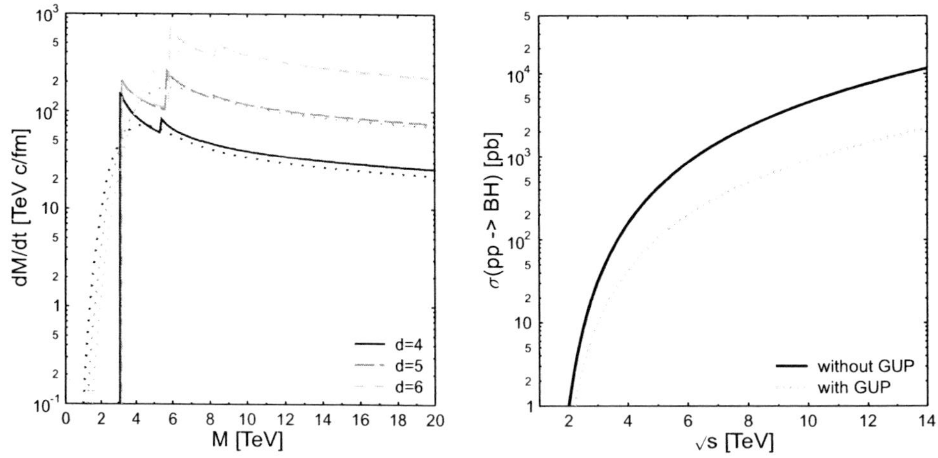

Figure 13: The left plot[66] shows the evaporation rate with discrete energy spectrum in comparison to the continuous case (dotted) for various d. The right plot, from [43], shows the total cross section for black hole production as a function of the center of mass energy $\sqrt{s}$ with and without the generalized uncertainty principle (GUP). For the expected LHC-energies the cross sections is lowered by about a factor 5. This result does barely depend on the number of extra dimensions.

has to be and thus, the energy of the lowest level increases. At black hole masses close to the Planck mass, it happens that even the energy of the lowest level would exceed the total energy of the black hole and can not be emitted.

The evaporation therefore will proceed in quantized steps and stops at a finite mass value. A relic is left.

The production of a relic instead of a final decay of the black hole would increase the missing p_T in the particle collision [67]. Also, a certain fraction of these objects will be charged which allows for a direct detection [68].

7 The Minimal Length

Gravity itself is inconsistent with physics at very short scales. The introduction of gravity into quantum field theory appears to spoil their renormalizability and leads to incurable divergences. It has therefore been suggested that gravity should lead to an effective cutoff in the ultraviolet, i.e. to a minimal observable length. It is amazing enough that all attempts towards a fundamental theory of everything necessarily seem to imply the existence of such a minimal length scale.

The occurrence of the minimal length scale has to be expected from very general reasons in all theories at high energies which attempt to include effects of quantum gravity. This can be understood by a phenomenological argument which has been examined in various ways. Test particles of a sufficiently high energy to resolve a distance as small as the Planck length, $l_{\mathrm{p}} = 1/m_{\mathrm{p}}$, are predicted to gravitationally curve and thereby to significantly disturb the very spacetime structure which they are meant to probe. Thus, in addition to the expected quantum uncertainty, there is another uncertainty caused which arises from spacetime fluctuations at the Planck scale.

Consider a test particle, described by a wave packet with a mean Compton-wavelength λ. Even in standard quantum mechanics, the particle suffers from an uncertainty in position Δx and momentum Δp, given by the standard Heisenberg-uncertainty relation $\Delta x \Delta p \geq 1/2$. Usually, every sample under investigation can be resolved by using beams of high enough momentum to focus the wave packet to a width below the size of the probe, Δx, which we wish to resolve. The smaller the sample, the higher the energy must become and thus, the bigger the collider.

General Relativity tells us that a particle with an energy-momentum $p \sim 1/\lambda$ in a volume of spacetime Δx^3 causes a fluctuation of the metric g. Einstein's Field Equations (1) allow us to relate the second derivative of the metric to the energy density and so we can estimate the perturbation with

$$\frac{\delta g}{\Delta x^2} \sim l_{\mathrm{p}}^2 \frac{p}{\Delta x^3} \quad . \tag{54}$$

This leads to an additional fluctuation, $\Delta x'$, of the spacetime coordinate frame of order $\Delta x' = \delta g \Delta x \approx l_{\mathrm{p}}^2 p$, which increases with momentum and becomes non-negligible at Planckian energies.

But from this expression we see not only that the fluctuations can no longer be neglected at Planckian energies and the uncertainty of the measurement is amplified. In addition, we see that a focussation of energy of the amount necessary to resolve the Planck scale leads to the formation of a black hole whose horizon is located at $\delta g \approx 1$, should the matter be located inside. Since the black hole radius has the property to expand linearly with the energy inside the horizon, both arguments lead to the same conclusion: a minimal uncertainty in measurement can not be erased by using test particles of higher energies. It is always $\Delta x > l_{\mathrm{p}}$: at low energies because the Compton-wavelengths are too big for a high resolution, at high energies because of strong curvature effects.

Thus, to first order the uncertainty principle is generalized to

$$\Delta x \gtrsim \frac{1}{\Delta p} + \text{const. } l_{\mathrm{p}}^2 p \quad . \tag{55}$$

Without doubt, the Planck scales marks a thresholds beyond which the old description of space-time breaks down and new phenomena have to appear.

More stringent motivations for the occurrence of a minimal length are manifold. A minimal length can be found in String Theory, Loop Quantum Gravity and Non-Commutative Geometries. It can be derived from various studies of thought-experiments, from black hole physics, the holographic principle and further more. Perhaps the most convincing argument, however, is that there seems to be no self-consistent way to avoid the occurrence of a minimal length scale. For reviews see e.g. [69].

Instead of finding evidence for the minimal scale as has been done in numerous studies, on can use its existence as a postulate and derive extensions to quantum theories with the purpose to examine the arising properties in an effective model.

Such approaches have been undergone and the analytical properties of the resulting theories have been investigated closely [70]. In the scenario without extra dimensions, the derived modifications are important mainly for structure formation and the early universe [71].

The importance to deal with a finite resolution of spacetime, however, is sensibly enhanced if we consider a spacetime with large extra dimensions. In this case, the new fundamental mass scale is lowered and thus, the minimal length is raised. The effects of a finite resolution of spacetime should thus become important in the same energy range in which the first effects of gravity are expected. Some consequences of the inclusion of the minimal length into the model of extra dimensions have been worked out in [72].

A generalized uncertainty principle, as follows from the existence of a minimal length scale, makes it increasingly difficult to focus the wave functions of the colliding particles. This is sketched in Figure 14. It requires higher energies to achieve an energy density inside the Schwarzschild radius which is sufficient for a collapse and the creation of a black hole.

The consequences of the generalized uncertainty for the total black hole cross section and the number of produced black holes have been derived in [43] and are depicted in Figure 13, right. As we expect, the cross section is lowered. For the estimated LHC-energies this results in a suppression of the black hole production by a factor ≈ 5.

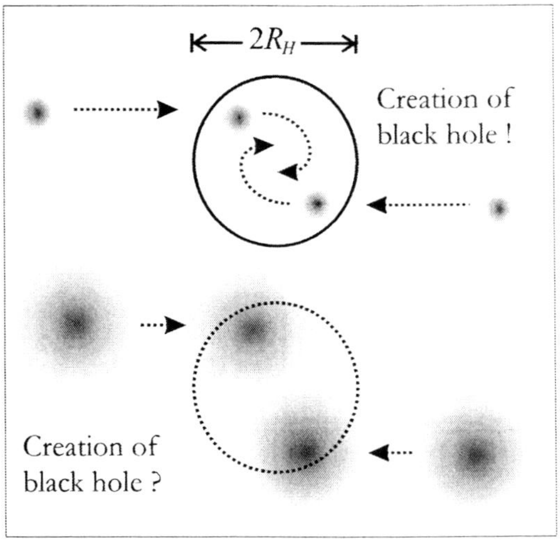

Figure 14: Schematic figure for parton scattering with and without generalized uncertainty.

The arising modifications at high energies do also influence the production of black holes. Using the generalized uncertainty principle, the properties [39] of the evaporation process are modified when the black hole radius approaches the minimal length which decreases the number of decay particles and leads to the formation of a stable remnant.

8 Are Those Black Holes Dangerous?

"Big Bang Machine: Will it destroy Earth?, Creation of a black hole on Long Island? The committee will also consider [...] that the colliding particles could achieve such a high density that they would form a mini black hole. In space, black holes are believed to generate intense gravitational fields that suck in all surrounding matter. The creation of one on Earth could be disastrous. [...]"

The London Times, July 18th, 1999

Considered the fact that black holes are known to swallow whole galaxy centers, the question might be raised whether the tiny black holes, produced in the laboratory, would also accrete mass, swallow it and grow until everything in reach would be vanished beyond the event horizon.

Fortunately, this question can without doubt be answered with 'No'. These earth produced black holes still have a very small gravitational attraction. An energy of some TeV might be huge for collider physics but for gravitational purposes it is tiny. The strength of the gravitational force, acting on electrons orbiting around a nucleus, is even in the presence of extra dimensions still by a factor $\approx 10^{20}$ weaker than the electromagnetic force[12].

We can estimate the possibility of black holes gaining mass based on our earlier given results. Looking at Figure 9, we see that the mass loss ratio, $\mathrm{d}M_-/\mathrm{d}t$ of the black hole is at least ≈ 1 TeV c/fm[13], for a large number of extra dimensions it would be even larger. As previously stated, the gravitational attraction of the black hole is negligible. Nevertheless, is the black hole placed in a medium which is considered to be extremely dense and hot in earthly terms, it will gain mass by particles crossing the horizon simply through thermal motion. This mass gain ratio can be estimated to be

$$\frac{dM_+}{dt} \sim \pi R_H^2 T_{\rm m}^4 \quad . \tag{56}$$

where $T_{\rm m}$ is the temperature of the medium and the fore-factor is the surface of the black hole as seen on our submanifold[14]. Inserting $\rho = T_{\rm m}^4$ to be the most dense medium produced on earth, e.g. a Quark Gluon Plasma, we have typically $\rho \sim 100$ MeV/fm^3. Using as a characteristic value $R_H = 1/M_{\rm f} \sim 1/\mathrm{TeV}$, we find that

$$\frac{\mathrm{d}M_+}{\mathrm{d}t} \sim 10^{-9}\frac{\mathrm{GeV}c}{\mathrm{fm}} \ll \frac{\mathrm{d}M_-}{\mathrm{d}t} \quad , \tag{57}$$

and so the black hole will loose mass much faster than it can gain mass.

Let us consider a further extreme case: a black hole, created by an ultra high energetic cosmic ray, which traverses a neutron star. Cosmic ray events can occur at energies as high as 10^8 TeV and thus, can yield black holes with an ultra high γ-factor, $\gamma \sim E/M \sim 10^7$. This

[12]The precise value depends on the number of extra dimensions.

[13]1 fm/c is $\approx 10^{-23}$ sec and is the typical timescale for high energy collisions.

[14]Note, that is not the total surface of the black hole as this is a $d+2$ dimensional hypersurface.

means, the black hole has an ultra high speed and due to the effects of Lorentz-contraction, it will see the medium of the neutron star in an even higher density. Or, to put it different, in the reference frame of the neutron star the black hole travels fast, undergoes the relativistic time dilation and evaporates slower.

In the reference frame of the black hole, the mass loss is still given by the above estimation but it now sees a medium with density $\gamma^2\rho$. The mass gain during passage of the neutron star is then increased by a factor $\gamma^2 \sim 10^{14}$. By multiplying Eq. (57) with this factor we see that in this case, the mass gain become comparable to the mass loss! The black hole thus has a chance to eat up a significant part of the neutron star. This process has been investigated in [73]. However, the estimation we have given here does only apply immediately after the creation of the black hole. Once it starts growing, it will become more massive and slow down. This in turn will decrease its γ-factor, the mass loss again becomes dominant and the black hole decays. A continuing growth is possible only in a very constrained parameter range which is already excluded [73].

PREJUDICES AGAINST BLACK HOLES

BLACK HOLES ARE MATHEMATICAL TRASH: Wrong! The existence of an event horizon is a perfectly well defined space time property. Moreover, their formation follows from such general assumptions that to present knowledge there is no way to avoid them. General Relativity *requires* black holes. Since the horizon radius is proportional to the mass M, it follows that the density needed for the horizon formation is proportional to M/M^3. This means, for large masses, the density can be arbitrarily small.

BLACK HOLES ARE SPECULATIVE: Wrong! The evidence for a black hole in the center of our galaxy (Sag. A*) is overwhelming [74]. More than 20 other candidates in our galaxy [75] and 10 further super massive black holes [76] have been established so far. The analysis of the motion of the surrounding objects is used to determine the amount mass concentrated in their vicinity. The mass inside the region is so high that the only plausible explanation is a black hole and *everything else* is highly speculative [77].

BLACK HOLES ARE BLACK: Wrong! The strong curvature effects in the vicinity of a black hole can rip apart virtual particle pairs and result in particle emission in the Hawking-radiation process. For objects of astrophysical sizes, this temperature is too low to be observable. These large black holes in general have rotating accretion disks surrounding them, in which the fast moving matter gets extremely hot and emits radiation.

BLACK HOLES ON EARTH ARE DANGEROUS: Wrong! The black holes which could possibly be created on earth would still have a very small mass and, connected with it, a very small gravitational attraction. Their interaction with other particles is so weak that they can not grow, even under extreme conditions, like inside a quark gluon plasma. In all cases, the evaporation processes will dominate, causing the black hole to decay instead to grow.

9 Summary: What Black Holes Can Teach Us

Black holes are a topic as fascinating as challenging. The investigations during the last century have led to important insights. Thermodynamics of black holes taught us of a deep connection between quantum theory, General Relativity and thermodynamics which has recently been studied further in the string/black hole correspondence and is a matter of ongoing research. Black holes and the puzzle of information loss have taught us about the possibility of stable black hole relics and the natural inclusion of a minimal length scale.

The presence of additional compactified dimensions would allow the production of tiny black holes in particle collisions. Their investigation in experiments on earth would provide a direct possibility to address the question of extra dimensions and the new fundamental scale. But moreover, it would allow us to probe the phenomenology of physics at the Planck scale and increase our knowledge about the underlying, yet to be found theory. Black holes can teach us to find the missing link between general relativity and quantum theory.

Acknowledgments

I want to thank Marcus Bleicher, Stefan Hofmann and Horst Stöcker for the inspiring discussions at Frankfurt University. I also thank Claudia Schumacher for critical remarks. I apologize for every missing citation on the topic. This work was supported by the German Academic Exchange Service (DAAD), NSF PHY/0301998 and the DFG.

A Useful Fomulae

$$\Gamma^{\alpha}{}_{\mu\nu} = \frac{1}{2}g^{\alpha\beta}(\partial_{\mu}g_{\beta\nu} + \partial_{\nu}g_{\beta\mu} - \partial_{\nu}g_{\mu\nu}) \tag{58}$$

$$R^{\mu}{}_{\nu\alpha\beta} = \partial_{\alpha}\Gamma^{\mu}{}_{\nu\beta} - \partial_{\alpha}\Gamma^{\mu}{}_{\nu\alpha} + \Gamma^{\mu}{}_{\rho\alpha}\Gamma^{\rho}{}_{\nu\beta} - \Gamma^{\mu}{}_{\rho\beta}\Gamma^{\rho}{}_{\nu\alpha} \tag{59}$$

$$R^{\mu}{}_{\nu} = R^{\alpha}{}_{\mu\alpha\nu} \tag{60}$$

$$R = g^{\mu\nu}R_{\mu\nu} \tag{61}$$

B Useful Constants

Planck length	$l_{\rm p}$	$\approx$	10^{-35} m	$\approx$	10^{-20} fm
Planck mass	$m_{\rm p}$	$\approx$	10^{-5} g	$\approx$	1.2×10^{16} TeV
Planck time	$t_{\rm p}$	$\approx$	10^{-43} s	$\approx$	10^{-20} fm/c
Mass of the sun	$M_{\odot}$	$\approx$	2×10^{33} g	$\approx$	2.4×10^{54} TeV
Radius of the sun	$R_{\odot}$	$\approx$	7×10^{8} m		
Schwarzschild radius of the Sun	R_H	$\approx$	3×10^{3} m		
c.o.m. energy at the LHC	$\sqrt{s}$			$\approx$	14 TeV

References

[1] K. Schwarzschild, "*Über das Gravitationsfeld eines Massenpunktes nach der Einsteinschen Theorie*" in "*Sitzungsberichte der Königlich Preussischen Akademie der Wissenschaften zu Berlin*", **189**-196 (1916).

[2] C. W. Misner, K. S. Thorne and J. A. Wheeler, "*Gravitation*", W. H. Freeman and Company (New York, 1973); R. U. Sexl and H. K. Urbantke, "*Gravitation Und Kosmologie*"", Spektrum Akademischer Verlag, 4. Auflage (1995).

[3] S. W. Hawking, *Comm. Math. Phys.* **43**, 199-220 (1975); *Phys. Rev. D* **14**, 2460-2473 (1976).

[4] R. M. Wald, " *General Relativity*", Chicago Lectures in Physics (Chicago, 1984).

[5] I. D. Novikov and V. P. Frolov, "*Black Hole Physics*", Kluver Academic Publishers (1998).

[6] N. D. Birell and P. C. W. Davies, "*Quantum Fields in Curved Space*", Cambridge University Press (Cambridge, 1982).

[7] R. Casadio and B. Harms, *Phys.Rev. D* **64**, 024016 (2001); *Phys.Lett. B* **487** 209-214 (2000).

[8] D. N. Page, *Phys. Rev. D* **13** 198 (1976).

[9] R. P. Kerr, *Phys. Rev. Lett.* **11** 237 (1963); G. Nordström, Proc. Kon. Ned. Akad. Wet. 20, 1238-1245, (1918); H. Reissner, Ann. Phys. 59, 106-120, (1916).

[10] P. Kanti, *Int. J. Mod. Phys. A* **19** (2004) 4899.

[11] I. Antoniadis, *Phys. Lett. B* **246**, 377 (1990).

[12] N. Arkani-Hamed, S. Dimopoulos and G. R. Dvali, *Phys. Lett. B* **429**, 263 (1998); I. Antoniadis, N. Arkani-Hamed, S. Dimopoulos and G. R. Dvali, *Phys. Lett. B* **436**, 257 (1998); N. Arkani-Hamed, S. Dimopoulos and G. R. Dvali, *Phys. Rev. D* **59**, 086004 (1999).

[13] L. Randall and R. Sundrum, *Phys. Rev. Lett.* **83**, 4690 (1999).

[14] L. Randall and R. Sundrum, *Phys. Rev. Lett.* **83**, 3370 (1999).

[15] T. Appelquist, H. C. Cheng and B. A. Dobrescu, *Phys. Rev. D* **64**, 035002 (2001); C. Macesanu, C. D. McMullen and S. Nandi, *Phys. Rev. D* **66**, 015009 (2002); T. G. Rizzo, *Phys. Rev. D* **64**, 095010 (2001).

[16] J. C. Long and J. C. Price, *Comptes Rendus Physique* **4**, 337 (2003); C. D. Hoyle *et al*, *Phys. Rev. D* **70**, 042004 (2004); C. D. Hoyle *et al*, *Phys. Rev. Lett.* **86**, 1418 (2001); J. Chiaverini *et al*, *Phys. Rev. Lett.* **90**, 151101 (2003).

[17] J. Hewett and M. Spiropulu, *Ann. Rev. Nucl. Part. Sci.* **52**, 397 (2002); T. G. Rizzo, *Phys. Rev. D* **64**, 095010 (2001); T. Han, J. D. Lykken and R. J. Zhang, *Phys. Rev. D* **59** (1999) 105006; S. Cullen, M. Perelstein and M. E. Peskin, *Phys. Rev. D* **62**, 055012 (2000).

[18] A. Goyal, A. Gupta and N. Mahajan, *Phys. Rev. D* **63**, 043003 (2001); R. Emparan, M. Masip and R. Rattazzi, *Phys. Rev. D* **65**, 064023 (2002); D. Kazanas and A. Nicolaidis, *Gen. Rel. Grav.* **35**, 1117 (2003).

[19] A. Ringwald and H. Tu, *Phys. Lett. B* **525** 135-142 (2002); J. Feng and A. Shapere, Phys. Rev. Lett. 88, 021303 (2002); A. Cafarella, C. Coriano and T. N. Tomaras, [arXiv:hep-ph/0410358]; L. A. Anchordoqui, J. L. Feng, H. Goldberg and A. D. Shapere, *Phys. Rev. D* **65** 124027 (2002); S. I. Dutta, M. H. Reno and I. Sarcevic, *Phys. Rev. D* **66**, 033002 (2002).

[20] K. Cheung, [arXiv:hep-ph/0409028]; G. Landsberg [CDF and D0 - Run II Collaboration], [arXiv:hep-ex/0412028].

[21] R. C. Myers and M. J. Perry *Ann. Phys.* **172**, 304-347 (1986).

[22] M. B. Voloshin, *Phys. Lett. B* **518**, 137 (2001); *Phys. Lett. B* **524**, 376 (2002); S. B. Giddings, in *Proc. of the APS/DPF/DPB Summer Study on the Future of Particle Physics (Snowmass 2001)* ed. N. Graf, eConf C010630, P328 (2001).

[23] S. N. Solodukhin, *Phys. Lett. B* **533**, 153 (2002); A. Jevicki and J. Thaler, *Phys. Rev. D* **66**, 024041 (2002); T. G. Rizzo, in *Proc. of the APS/DPF/DPB Summer Study on the Future of Particle Physics (Snowmass 2001)* ed. N. Graf, eConf C010630, P339 (2001); D. M. Eardley and S. B. Giddings, *Phys. Rev. D* **66**, 044011 (2002).

[24] V. S. Rychkov, *Phys. Rev. D* **70**, 044003 (2004); K. Kang and H. Nastase, [arXiv:hep-th/0409099].

[25] I. Ya. Yref'eva, Part.Nucl. 31, 169-180 (2000). S. B. Giddings and V. S. Rychkov, *Phys. Rev. D* **70**, 104026 (2004); V. S. Rychkov, [arXiv:hep-th/0410041]; T. Banks and W. Fischler, [arXiv:hep-th/9906038]. O. V. Kancheli, [arXiv:hep-ph/0208021].

[26] H. Yoshino and Y. Nambu, *Phys. Rev. D* **67**, 024009 (2003).

[27] G. T. Horowitz and J. Polchinski, *Phys. Rev. D* **66** 103512 (2002).

[28] S. N. Solodukhin, *Phys. Lett. B* **533** 153-161 (2002); D. Ida, K. y. Oda and S. C. Park, *Phys. Rev. D* **67**, 064025 (2003).

[29] R. Casadio and B. Harms, *Int. J. Mod. Phys. A* **17**, 4635 (2002);

[30] P. Burikham, T. Figy and T. Han, [arXiv:hep-ph/0411094].

[31] S. Dimopoulos and R. Emparan, *Phys. Lett. B* **526**, 393 (2002).

[32] G. T. Horowitz and J. Polchinski, *Phys. Rev. D* **55**, 6189 (1997); T. Damour and G. Veneziano, *Nucl. Phys. B* **568**, 93 (2000); E. Halyo, B. Kol, A. Rajaraman and L. Susskind, *Phys. Lett. B* **401**, 15 (1997); L. Susskind, [arXiv:hep-th/9309145].

[33] G. Veneziano, *JHEP* **0411**, 001 (2004), [arXiv:hep-th/0410166].

[34] S. Dimopoulos and G. Landsberg *Phys. Rev. Lett.* **87**, 161602 (2001); P.C. Argyres, S. Dimopoulos, and J. March-Russell, *Phys. Lett. B* **441**, 96 (1998).

[35] R. Emparan, G. T. Horowitz and R. C. Myers *Phys. Rev. Lett.* **85**, 499 (2000).

[36] S. B. Giddings and S. Thomas, *Phys. Rev. D* **65** 056010 (2002).

[37] K. m. Cheung, *Phys. Rev. Lett.* **88**, 221602 (2002); Y. Uehara, [arXiv:hep-ph/0205068]; *Prog. Theor. Phys.* **107**, 621 (2002); L. Anchordoqui and H. Goldberg, *Phys. Rev. D* **67**, 064010 (2003); M. Bleicher, S. Hofmann, S. Hossenfelder and H. Stöcker, *Phys. Lett.* **548**, 73 (2002); J. Alvarez-Muniz, J. L. Feng, F. Halzen, T. Han and D. Hooper, *Phys. Rev. D* **65**, 124015 (2002). I. Mocioiu, Y. Nara and I. Sarcevic, *Phys. Lett. B* **557**, 87 (2003).

[38] S. Hossenfelder, S. Hofmann, M. Bleicher and H. Stöcker, *Phys. Rev. D* **66**, 101502 (2002).

[39] M. Cavaglia, S. Das and R. Maartens, *Class. Quant. Grav.* **20**, L205 (2003); M. Cavaglia and S. Das, *Class. Quant. Grav.* **21**, 4511 (2004).

[40] J. Tanaka, T. Yamamura, S. Asai and J. Kanzaki, [arXiv:hep-ph/0411095].

[41] C. M. Harris, M. J. Palmer, M. A. Parker, P. Richardson, A. Sabetfakhri and B. R. Webber, JHEP **0505**, 053 (2005) [arXiv:hep-ph/0411022].

[42] R. Godang, S. Bracker, M. Cavaglia, L. Cremaldi, D. Summers and D. Cline, [arXiv:hep-ph/0411248].

[43] S. Hossenfelder, *Phys. Lett. B* **598**, 92 (2004).

[44] G. Altarelli and G. Parisi, *Nucl. Phys. B* **126**, 298 (1977); Y. L. Dokshitzer, *Sov. Phys. JETP* **46**, 641 (1977); L. N. Lipatov, *Sov. J. Nucl. Phys.* **20**, 94 (1975); V. N. Gribov and L. N. Lipatov, *Yad. Fiz.* **15**, 1218 (1972).

[45] K. R. Dienes, E. Dudas and T. Gherghetta, *Nucl. Phys. B* **537**, 47 (1999); *Phys. Lett. B* **436**, 55 (1998).

[46] G. Landsberg, [arXiv:hep-ph/0211043]; M. Cavaglia, *Int. J. Mod. Phys. A* **18**, 1843 (2003).

[47] T. Sjostrand, L. Lonnblad and S. Mrenna, [arXiv:hep-ph/0108264].

[48] C. M. Harris, P. Richardson and B. R. Webber, *JHEP* **0308**, 033 (2003) [arXiv:hep-ph/0307305].

[49] S. Hawking, *Commun. Math. Phys.* **87**,395 (1982).

[50] S. Chandrasekhar, "*The Mathematical Theory of Black Holes*", Oxford University Press, (Oxford 1983).

[51] J. Preskill, [arXiv:hep-th/9209058].

[52] D. N. Page, *Phys. Rev. Lett.* **44**, 301 (1980); G. t'Hooft, *Nucl. Phys. B* **256** 727 (1985); A. Mikovic, *Phys. Lett.* **304 B**, 70 (1992); E. Verlinde and H. Verlinde, *Nucl. Phys.* **B 406**, 43 (1993); L. Susskind, L. Thorlacius and J. Uglum, *Phys. Rev. D* **48**, 3743 (1993); D. N. Page, *Phys. Rev. Lett.* **71** 3743 (1993).

[53] Y. Aharonov, A. Casher and S. Nussinov, *Phys. Lett.* **191** ,B 51 (1987); T. Banks, A. Dabholkar, M. R. Douglas and M. O'Loughlin, *Phys. Rev. D***45** 3607 (1992); T. Banks and M. O'Loughlin, *Phys. Rev. D* **47**, 540 (1993); T. Banks, M. O'Loughlin and A. Strominger, *Phys. Rev. D***47**, 4476 (1993); S. B. Giddings, *Phys. Rev. D* **49**, 947 (1994).

[54] M. A. Markov, in: "*Proc. 2nd Seminar in Quantum Gravity*", edited by M. A. Markov and P. C. West, Plenum, New York (1984).

[55] Y. B. Zel'dovich, in: "*Proc. 2nd Seminar in Quantum Gravity*", edited by M. A. Markov and P. C. West, Plenum, New York (1984).

[56] J. D. Barrow, E. J. Copeland and A. R. Liddle, *Phys. Rev. D* **46**, 645 (1992).

[57] B. Whitt, *Phys. Rev. D* **38**, 3000 (1988).

[58] R. C. Myers and J. Z. Simon, *Phys. Rev. D* **38**, 2434 (1988).

[59] C. G. Callan, R. C. Myers and M. J. Perry, *Nucl. Phys. B* **311**, 673 (1988).

[60] S. Alexeyev, A. Barrau, G. Boudoul, O. Khovanskaya and M. Sazhin, *Class. Quant. Grav.* **19**, 4431-4444 (2002).

[61] M. J. Bowick, S. B. Giddings, J. A. Harvey, G. T. Horowitz and A. Strominger, Phys. Rev. Lett. **61** 2823 (1988).

[62] S. Coleman, J. Preskill and F. Wilczek, *Mod. Phys. Lett.A* **6** 1631 (1991).

[63] K-Y. Lee, E. D. Nair and E. Weinberg, *Phys. Rev. Lett.* **68** 1100 (1992).

[64] G. W. Gibbons and K. Maeda, *Nucl. Phys. B* **298** 741 (1988).

[65] T. Torii and K. Maeda, *Phys. Rev. D* **48** 1643 (1993).

[66] S. Hossenfelder, M. Bleicher, S. Hofmann, H. Stocker and A. V. Kotwal, *Phys. Lett. B* **566**, 233 (2003).

[67] B. Koch, M. Bleicher and S. Hossenfelder, [arXiv:hep-ph/0507138] (accepted for publication in JHEP).

[68] S. Hossenfelder, B. Koch and M. Bleicher, [arXiv:hep-ph/0507140].

[69] L. J. Garay, *Int. J. Mod. Phys. A* **10**, 145 (1995); Y. J. Ng, *Mod. Phys. Lett. A* **18**, 1073 (2003); S. Hossenfelder, *Mod. Phys. Lett. A* **19**, 2727 (2004).

[70] A. Kempf, G. Mangano and R. B. Mann, *Phys. Rev. D* **52**, 1108 (1995); M. Maggiore, *Phys. Lett. B* **319**, 83 (1993); M. Maggiore, *Phys. Rev. D* **49**, 5182 (1994); A. Camacho, *Int. J. Mod. Phys. D* **12**, 1687 (2003); I. Dadic, L. Jonke and S. Meljanac, *Phys. Rev. D* **67** (2003) 087701; F. Brau, *J. Phys. A* **32** (1999) 7691; R. Akhoury and Y. P. Yao, Phys. Lett. B **572**, 37 (2003).

[71] U. H. Danielsson, *Phys. Rev. D* **66** (2002) 023511; S. Shankaranarayanan, *Class. Quant. Grav.* **20** (2003) 75; L. Mersini, M. Bastero-Gil and P. Kanti, *Phys. Rev. D* **64** (2001) 043508; A. Kempf, *Phys. Rev. D* **63** (2001) 083514; A. Kempf and J. C. Niemeyer, *Phys. Rev. D* **64** (2001) 103501; J. Martin and R. H. Brandenberger, *Phys. Rev. D* **63** (2001) 123501; R. Easther, B. R. Greene, W. H. Kinney and G. Shiu, R. H. Brandenberger and J. Martin, *Mod. Phys. Lett. A* **16** (2001) 999; R. Casadio and L. Mersini, *Int. J. Mod. Phys. A* **19**, 1395 (2004); G. L. Alberghi, R. Casadio and A. Tronconi, *Phys. Lett. B* **579**, 1 (2004).

[72] G. Bhattacharyya, P. Mathews, K. Rao and K. Sridhar, [arXiv:hep-ph/0408295]. S. Hossenfelder, M. Bleicher, S. Hofmann, J. Ruppert, S. Scherer and H. Stocker, *Phys. Lett. B* **575**, 85 (2003). S. Hossenfelder, *Phys. Rev. D* **70**, 105003 (2004).

[73] P. Gorham, J. Learned and N. Lehtinen, [arXiv:astro-ph/0205170].

[74] L. Ferrarese and H. Ford, [arXiv:astro-ph/0411247].

[75] J. E. McClintock and R. A. Remillard, [arXiv:astro-ph/0306213].

[76] J. Kormendy and K. Gebhardt, [arXiv:astro-ph/0105230].

[77] J. E. McClintock, R. Narayan and G. B. Rybicki, [arXiv:astro-ph/0403251]; R. Narayan, [arXiv:astro-ph/0310692]; H. Falcke, F. Melia and E. Agol, [arXiv:astro-ph/9912263].

In: Focus on Black Hole Research
Editor: Paul V. Kreitler, pp. 159-179
ISBN 1-59454-460-3

Chapter 5

COSMOLOGY WITH PRIMORDIAL BLACK HOLES

A. S. Majumdar*
S. N. Bose National Centre for Basic Sciences, Block JD,
Sector III, Salt Lake, Calcutta 700098, India

Abstract

We consider the evolution of primordial black holes formed due to the gravitational collapse of overdense regions in the early universe. These black holes are formed with an initial mass spectrum, and lose energy by the process of Hawking evaporation. However, the process of accretion of radiation in the radiation dominated era could under certain conditions counterbalance the effect of evaporation, and this leads to longer lifetime for the black holes. We investigate the scenarios of the early universe with a population of primordial black holes interacting with the surrounding radiation in standard cosmology and as well as in braneworld cosmology separately. In the braneworld scenario the 5-dimensonal black hole geometry leads to modified evaporation and efficient accretion that is particularly conducive to the survival of primordial black holes up to several cosmologically interesting eras. Primordial black holes in the braneworld scenario could act as candidates for non-baryonic cold dark matter. A mechanism for the formation of binaries with primordial braneworld black holes is also discussed.

1 Introduction

Primordial black holes have numerous implications on processes occurring during diverse cosmological epochs. Black holes could be formed in the early universe through a variety of mechanisms[1]. A widely studied mechanism is that of the cosmological density perturbations generated through inflation[2]. Inflation leads to a scale invariant spectrum of primordial density perturbations which tightly constrains the production rate and the initial mass

*E-mail address: archan@bose.res.in

spectrum of primordial black holes formed through the collapse of density perturbations[3]. Recent observational results on the Cosmic Microwave Background Radiation[4] favoring a primordial power spectrum with a spectral index very close to one ($n_s \simeq 1$) have motivated the detailed investigation of the production of primordial black holes in inflationary models with distinguished scales or broken scale invariance[5].

Due to the fact that black holes can accrete their surrounding energy as well as radiate energy through the process of Hawking evaporation, the interest in the study of the evolution of primordial black holes[6] ranges over a very big time span starting from their formation and running right through their lifetimes which, in certain cases could even exceed the present age of the universe. A key question concerning primordial black holes is their longevity which depends crucially on the effectiveness of various accretion processes. Upon formation in the early universe, the black holes are surrounded by radiation which is the dominant form of available energy at that time. The debate as to how significant the process of accretion of radiation is for primordial black holes in the standard cosmological scenario, ranges till date[6, 7], though numerical simulations have indicated considerable increase of black holes lifetimes with notable consequences[8, 9]. Primordial black holes could also accrete the energy of background scalar fields that are present in a variety of scalar-tensor models in cosmology[11]. If the present day acceleration of the universe results from a dominant scalar field energy, as in models of quintessence[12], then primordial black holes could grow up enormously in size by accreting such dark energy. Indeed, the observational evidence in the favor of the abundance of supermassive black holes in galactic centres[13] presents a challenge for theoretical astrophysicists to formulate formation mechanisms for such huge black holes, and primordial black holes could be potential candidates[14].

Most of the studies concerning the implications of primordial black holes on cosmology have focussed on the products emanating from Hawking radiation. If primodial black holes are formed with a spectrum of initial masses, they are likely to have a range of lifetimes corresponding to different epochs of the universe. During their later stages of evolution, the Hawking temperature exceeds the threshold of heavy particle production. It was realized that the baryon number violating and CP-violating decay of heavy particles radiated by them could lead to baryogenesis[15]. If a fraction of the population of primordial black holes have lifetimes of the order of or somewhat exceeding the electroweak phase transition time, then the baryons produced would evade the sphaleron wash-out[16]. This mechanism could be responsible for generating the observed baryon asymmetry in the universe[8, 10]. But, in general the decay of primordial black holes would produce an excess of light particles and photons. Nucleosynthesis results can therefore be used to constrain the abundance of primordial black holes in the relevant mass range[17]. Further bounds on the primordial black hole abundance comes through the Hawking emission of photons that could distort the CMBR spectrum at high frequencies[18]. A comprehensive discussion of the constraints on the primordial black holes abundances imposed by observational requirements at various cosmological epochs can be found in Ref.[19].

An exciting possibility featuring primordial black holes is their capacity to be a significant portion of cold dark matter. It has been claimed recently that a scenario with primordial black holes as cold dark matter could be viable for particular models of broken scale invariance for the primordial power spectrum[20]. Primordial black holes also produce isocurvature perturbations with interesting consequences[21]. The probe of black holes as dark compact objects in galaxy haloes is the target of much recent observational activity[22]. A large number of these black holes are expected to be in the form of binaries. Different observational evidences as well as theoretical arguments ranging from lensing effects to hypervelocity star ejections have been put forward to bolster the claim in support of the existence of black hole binaries[23] in a wide mass spectrum. Recently, it has been proposed that sub-lunar mass binaries comprised of braneworld black holes could produce detectable gravitational waves in their coalescing stage[24]. Gravitational lensing experiments do indeed leave open the possibility of the existence of sub-lunar mass black holes in several mass ranges[25].

Primordial black holes have been studied extensively in the standard cosmological model, though their remain several gaps in our understanding of their formation, evolution and implications. Recent resurgence of interest in the physics of primordial black holes comes also from the advent of braneworld models of the universe. Indeed, the entire scenario of cosmology with primordial black holes gets modified in the context of braneworld gravity. There has been widespread activity in braneworld gravity recently[26]. The braneworld scenario of our universe as enunciated in the Randall-Sundrum (RS-II) model[27] opens up the fascinating possibility of the existence of a large extra spatial dimension. Consistency of the RS-II model demands that the standard model fields are confined to the brane, except for gravity which could also propagate into the bulk. This is ensured by the positive tension brane, and the Ads5 bulk with a negative cosmological constant and curvature radius l. The resultant modification of the Newtonian gravitational potential is being probed by current experiments which have set the upper limit of the scale of the extra dimension in the sub-millimeter region, i.e., $l \leq 0.2\mathrm{m}m$[28]. Standard Schwarzschild black hole solutions have to be modified in the context of this model[29, 30]. Furthermore, the evolution of primordial black holes and their multifarious consequences require completely new investigation in braneworld cosmology.

In this article we will first present a review of the general features of cosmology with a population of primordial black holes. This will be done in Section 2 completely within the context of standard cosmology, with the aim of highlighting the essential features of primordial black hole evolution taking into account accretion and Hawking evaporation. Although we will explicitly consider accretion of radiation in the radiation dominated era of the universe, a similar formalism could also be applied to the case of accretion of other kinds of energy like that of a background scalar field[11, 14]. We will next move on to the braneworld scenario. In Section 3 we will discuss the evolution of primordial black holes formed during the early braneworld era. We will highlight the fact that accretion of radiation is a significant effect for braneworld black holes, which leads to mass growth and prolonged lifetimes for the primordial population. Primordial black holes in the braneworld scenario have potentially rich consequences, many of which are yet to be investigated in detail. In Section 4 we will present a somewhat detailed analysis of one feature of the braneworld scenario with primordial black holes, viz., the seeding of black hole binaries in the early braneworld era. In Section 5 we conclude with a summary of the key differences of the braneworld black hole scenario compared to the cosmology with standard black holes.

2 Primordial Black Holes in Standard Cosmology

We begin by considering a population of primordial black holes in the early radiation dominated era of the universe. As stated in the introduction these black holes could be formed by various mechanisms[1] operative in the post-inflationary stage of the universe. In the simplest analysis we do not take into account effects of the initial mass distribution[3, 5] of black holes, but work with the uniform black holes mass M. The time evolution of the FRW scale factor $a(t)$ of the universe is governed by the equation

$$\frac{\dot{a}^2}{a^2} = \frac{8\pi G}{3}(M(t)n_{BH}(t) + \rho_R(t)) \tag{1}$$

where $n_{BH}(t)$ is the number density of the black holes at the epoch t, and $\rho_R(t)$ is the energy density of radiation. The number density of black holes scales as a^{-3}, so that

$$n_{BH}(t) = \frac{N_{BH}}{a^3(t)} \tag{2}$$

where the total number of black holes formed at time t_0 is $N_{BH} = n_{BH}(t_0)a^3(t_0)$.

The mass of a black hole will change with time due to two factors. First, the process of Hawking radiation causes a loss of mass. Secondly, accretion of the surrounding matter by the black hole leads to an increase in its mass. The rate of change of mass of a single black hole can be written as

$$\dot{M} = 4\pi R_{BH}^2 \Big(-\sigma T_{BH}^4 + f\rho_R\Big) \tag{3}$$

where R_{BH} is the Schwarzschild radius, σ is the Stefan-Boltzmann constant, T_{BH} is the Hawking temperature, and f is an $O(1)$ accretion efficiency factor. Relativistic particles and photons surrounding the black hole dominate the energy density of the universe. The thermalysed plasma has density ρ_R and a temperature T given by

$$\rho_R = \frac{\pi^2 g T^4}{30} \tag{4}$$

where g is the number of relativistic particle species. The first term on the right-hand side of Eq.(3) corresponds to the Hawking radiation effect with Hawking temperature given by

$$T_{BH} = \frac{M_4^2}{8\pi M} \tag{5}$$

with M_4 denoting the Planck mass. The second term in Eq.(3) corresponds to the accretion rate of the surrounding radiation ρ_R which during the radiation dominated era falls off as

$1/t^2$. Thus, the rate equation for an isolated black hole in the radiation dominated era can be written as

$$\dot{M} = -\frac{A_s M_4^4}{M^2} + \frac{B_s M^2}{M_4^2 t^2} \tag{6}$$

where A_s and B_s are dimensionless numbers[8, 9].

The cross-section of the capture of relativistic matter by the black hole could be somewhat different from the geometrical cross section $4\pi R_{BH}^2$ due to the effects of gravitational deflection of he surrounding matter towards the black hole and the finite mean free path of particles being captured. The size of the hubble radius which grows as $R_H \propto t$ limits the amount of surrounding energy that could be causally accreted by the black hole, thereby affecting the rate of accretion for black holes of the similar size. The actual capture cross-section also depends on the relative motion of the black hole with respect to the cosmic frame. A black hole moving with velocity v sees the background radiation temperature T to have an angular dependence

$$T(\theta) = \frac{T}{\gamma(1 - v cos\theta)} \tag{7}$$

with the Lorentz γ factor impeding Hawking radiation for black holes moving with relativistic velocities. This implies that in the cosmic frame the rate at which the mass of the black hole changes is[8]

$$\dot{M} = 4\pi R_{BH}^2 \sigma \left(-\frac{T_{BH}^4}{\gamma} + \frac{1}{3}(4\gamma - \frac{1}{\gamma})T^4 \right) \tag{8}$$

A high value of γ will enhance the mass gain due to accretion and also inhibit mass loss due to Hawking radiation, leading to longer life for the black hole[8]. The process of accretion entails several different complexities[6, 7, 8, 9]. For the sake keeping the present analysis as uncomplicated as possible, we will work with the unknown accretion efficiency factor f introduced in Eq.(3).

The interplay of accretion and Hawking radiation for a black hole balance at a particular mass M_c. Setting $\dot{M} = 0$ in Eq.(3), one can obtain M_c using Eqs.(4) and (5) to be

$$M_c = \frac{\sigma M_4^2}{8\pi (f\rho_R)^{1/4}} \tag{9}$$

If a black hole is formed with an initial mass lesser than M_c, it will be short-lived as a result of the Hawking evaporation process, the rate for which will keep increasing with time. Accretion will never be able to keep pace with this runaway evaporation. However, a black hole formed with an initial mass $M_0 > M_c$ will keep growing due to accretion of radiation till enough radiation density is available. The critical mass M_c corresponds to the Hawking temperature $(T_{BH})_c$ given by

$$(T_{BH})_c = \frac{\pi^2 f g}{30\sigma} T \tag{10}$$

Thus a black hole with Hawking temperature lesser than $(T_{BH})_c$, will be in an accreting phase. As stated earlier, the duration of this accreting phase will depend on the energy density of radiation which in turn is affected by the number density of black holes present.

The energy density of the relativistic matter ρ_R will evolve both due to the expansion of the universe, as well as because of the absorption or emission by the black holes. The time-variation of ρ_R is thus given by

$$\frac{d}{dt}(\rho_R(t)a^4(t)) = -\dot{M}N_{BH}a(t) \tag{11}$$

When the right-hand side of Eq.(3) vanishes, it yields the usual result $\rho_R \propto a^{-4}$. This happens when either no black holes are formed, i.e., $N_{BH} = 0$, or when the individual black hole masses do not change with the cosmic time.

In our simplistic analysis let us assume that at time t_0 a certain fraction β of the total energy density of the universe is trapped in N_{BH} number of black holes, all of which have approximately the same mass M_0. The remaining fraction $(1-\beta)$ of the energy goes into relativistic particles. We assume a small value of β corresponding to a radiation dominated era at t_0. Denoting ρ_{BH} as the sum of the energy densities of all the black holes, one has

$$\rho(t_0) = \rho_R(t_0) + \rho_{BH}(t_0) \tag{12}$$

$$\rho_R(t_0) = (1-\beta)\rho(t_0) = \frac{\pi^2 g}{30}T^4 \tag{13}$$

$$\rho_{BH}(t_0) = \beta\rho(t_0) = M_0 n_{BH}(t_0) \tag{14}$$

Let us now write down together the equations governing the cosmological evolution of the universe beyond the time t_0 given by the hubble expansion Eq.(1), the evolution of radiation density Eq.(11), and the rate of change of black hole mass Eq.(3), using Eqs.(12–14), i.e.,

$$\frac{\dot{a}^2}{a^2} = \frac{8\pi}{3M_4^2}\left(\rho_R + \frac{M\beta\rho(t_0)}{M_0}\frac{a^3(t_0)}{a^3}\right) \tag{15}$$

$$\dot{\rho_R} = -4\rho_R\frac{\dot{a}}{a} - \frac{\dot{M}\beta\rho(t_0)}{M_0}\frac{a^3(t_0)}{a^3} \tag{16}$$

$$\dot{M} = -\frac{A_s M_4^4}{M^2} + \frac{16\pi f M^2 \rho_R}{M_4^4} \tag{17}$$

The above coupled Eqs.(15),(16) and (17) can be numerically integrated for relevant initial conditions at t_0. This analysis has been performed by several authors[8, 9, 15]. The generic results can be summarized as follows. If one chooses a sufficiently small value of β, and the initial black hole masses greater than the critical mass M_c, then accretion of radiation is the dominant effect initially. This causes the energy density of radiation to decrease rapidly. Hence the temperature of the universe falls down at a faster rate than that arising out of the usual expansion of the scale factor. (For larger values of β, the universe may enter an early and non-standard matter (black hole) dominated epoch. During such a phase accretion by the black holes becomes increasingly insignificant because of very rapidly falling radiation density. We will describe such an early black hole dominated expansion in some details within the context of braneworld cosmology in Section 3). Though the accretion of radiation is unable to increase substantially the mass of the black holes (unless one has an extremely small β), it can cause considerable increase of black hole lifetime. Depending on the initial mass M_0, at some stage the black holes enter a quiescent phase where their mass stays nearly constant since both the accretion and the evaporation rates are negligible. During this stage the rate of hubble expansion is slightly increased compared to the standard radiation dominated expansion. The black hole evaporation rate then slowly begins to rise and this causes reheating of the universe. Towards the end of their lifetimes the black holes emit heavy particles which could produce the baryon asymmetry of the universe[8, 9, 15, 10]. However, if the black holes are formed with suitably larger initial mass, they could survive up to later eras, possibly even contributing a fraction to the cold dark matter density of the universe[20].

The initial mass M_0 of course determines the actual cosmological epoch when the black holes evaporate out. It is expected that black holes are produced with an initial mass spectrum in any realistic scenario of black hole formation[2, 3, 5]. A mass distribution could introduce interesting peculiarities into the dynamics at particular epochs over and above the simplified scenario considered by us. Heavier black holes that are formed later are likely to hinder accretion of the smaller one produced earlier. (This specific feature will be discussed in the context of braneworld black holes in Section 4). A distribution function could be defined in general quantifying the mass and energy fraction in black holes formed at different times. The effect of a mass distribution can be incorporated into the cosmological evolution (Eqs.(15),(16) and (17)) by introducing an additional differential equation for the distribution function and specifying further initial conditions related to the distribution function (see Ref.[9] for details). Of course, the mass distribution should be such as to ensure that the closure of the total energy density has not taken place at any epoch. This consideration leads to bounds on the initial mass spectrum. Moreover, since the black holes radiate most of their energies at the very end of their lifetimes, cosmological processes operating at that particular epoch are affected. Therefore known observations can be used to impose further bounds on the initial distribution. A number of such analyses have been carried out[19], but without taking into account the process of accretion. The accretion of radiation by primordial black holes affects the dynamics of cosmological evolution even more significantly in the braneworld context, as we will see in the next section.

3 Braneworld Cosmology with Black Holes

The cosmology of the RS-II model entails a modified high energy phase in the early radiation dominated era of the universe during which the right hand side of the Einstein equation contains terms that are quadratic in the brane energy momentum tensor[26]. Other modifications include the so-called "dark-energy" term which is given by the projection of the bulk Weyl tensor. Transition to the standard radiation dominated era takes place when $t >> t_c \equiv l/2$. Such a modified high energy evolution has rich consequences for the physics of the early universe[31]. In particular, the inflationary scenario is altered, allowing the possibility of steep inflaton potentials to accomplish the desired features[32]. Constraints on the duration of the brane dominated high energy phase are enforced by the necessity of conforming to the standard cosmological observational features such as nucleosynthesis and density perturbations.

Let us first review briefly some of the important features of the RS-II braneworld cosmology. The effective 4-dimensional Einstein tensor on the brane is given by[26]

$$G_{\mu\nu} = \frac{8\pi}{M_4^2}\tau_{\mu\nu} + \kappa^4\Pi_{\mu\nu} - E_{\mu\nu} \tag{18}$$

where $\tau_{\mu\nu}$ is the brane energy-momentum tensor; $\Pi_{\mu\nu}$ is quadratic in the brane EM tensor; and $E_{\mu\nu}$ is the projection of the 5-d Weyl tensor. The 4-dimensional Planck's mass M_4 is related to the gravitational coupling constant κ and the AdS length l by $\frac{8\pi}{M_4^2} = \frac{\kappa^2}{l}$. For the Friedmann-Robertson-Walker metric on the brane, the Friedmann equation is given by

$$H^2 = \frac{8\pi}{3M_4^2}\left(\rho + \frac{\rho^2}{2\lambda} + \rho_{KK}\right) + \frac{\Lambda_4}{3} - \frac{k}{a^2} \tag{19}$$

with H being the Hubble constant, ρ the energy density, and $k = -1, 0, 1$ representing open, flat and closed branes, respectively. ρ_{KK} is the effective energy density coming from the bulk Weyl tensor, $\lambda \equiv 3M_5^6/4\pi M_4^2$ is the brane tension, and Λ_4 the effective 4-d cosmological constant on the brane. The AdS curvature radius l is given by the bulk cosmological constant Λ_5 and 5-d Planck mass M_5 as $\Lambda_5 = -(3M_5^3)/(4\pi l^2)$. The induced 4-d cosmological constant Λ_4 is given by

$$\Lambda_4 = 3\left(\frac{M_5^6}{M_4^4} - \frac{1}{l^2}\right) \tag{20}$$

Setting $\Lambda_4 = 0$, one obtains a relation between the brane tension and AdS radius given by

$$\lambda^{-1/4} = \left(\frac{4\pi}{3}\right)^{1/4}\left(\frac{l}{l_4}\right)^{1/2} l_4 \tag{21}$$

Nucleosynthesis constrains the "dark energy" term ρ_{KK} to be negligible compared to the radiation density ρ[33]. In the following analysis we will neglect ρ_{KK}.

Assuming a radiation dominated equation of state, the $(k = 0)$ solutions for the Friedmann equation are given by

$$\rho_R = \frac{3M_4^2}{32\pi t(t + t_c)} \tag{22}$$

for the energy density, and

$$a = a_0\left[\frac{t(t + t_c)}{t_0(t_0 + t_c)}\right]^{1/4} \tag{23}$$

for the scale factor a during the radiation dominated era, and where $t_c \equiv l/2$ effectively demarcates the brane dominated "high energy" era from the standard radiation dominated era. For times earlier that t_c, i.e., $t \leq t_c$ (or $\rho \geq \lambda$), one has the non-standard high energy regime during which the radiation density and the scale factor evolve as

$$\rho_R = \frac{3M_4^2}{32\pi t_c t} \tag{24}$$

and

$$a = a_0 \left(\frac{t}{t_0}\right)^{1/4} \tag{25}$$

respectively. As a consequence, the time-temperature relation also gets modified during the brane dominated high energy era, i.e., $T \propto t^{-1/4}$. But, for times much later than t_c, i.e., $t >> t_c$ (or $\rho << \lambda$), one gets back the standard radiation dominated cosmological evolution given by

$$\rho_R = \frac{3M_4^2}{32\pi t^2} \tag{26}$$

and

$$a = a_0 \left(\frac{t}{(t_0 t_c)^{1/2}}\right)^{1/2} \tag{27}$$

Our purpose here is to explore the evolution of primordial black holes in this non-standard braneworld scenario.

The formation and evolution of black holes is an interesting and complex issue of investigation in braneworld cosmology. Horizon sized density perturbations in the high energy phase could lead to the formation of black holes on the brane by the process of gravitational collapse. The geometry of such black holes is inherently 5-dimensional[34]. There have been several recent efforts on understanding different types of black hole solutions on the brane[29]. In particular, black holes that are formed out of the collapse of horizon sized density perturbations, have been argued to have a 5-dimensional Schwarzschild metric, which when projected on the brane takes the form of a modified 4-dimensional Schwarzschild solution[30]. The horizon radii of such black holes is proportional to the square root of their masses, a feature that modifies the Hawking temperature, and consequently slows down the evaporation process[30]. Note however, that the behaviour of large black holes that are super-horizon sized having radius much greater than the Ads curvature radius could be completely different inasmuch as they may evaporate out rapidly as a consequence of Ads-CFT correspondence[35]. We will restrict our attention here to small sub-horizon sized black holes only. Since the 5-dimensional fundamental scale could be several orders of magnitude below the Planck scale, a lot of current excitement stems from the possibility of braneworld black holes being produced in high energy particle collisions either in the laboratory, or in cosmic ray showers[36]. Such processes could also be responsible for producing primordial black holes out of the thermal plasma in the early universe. In this article, we will not explore the details of any particular mechanism of primordial black hole production. In fact, a comprehensive analysis of the formation of primordial black holes in the braneworld scenario remains to be performed. Our analysis here deals with the evolution of black holes once they are formed in the high energy braneworld regime.

To this end, let us first consider the evolution of a single primordial black hole which is formed with a sub-horizon mass in the high energy radiation dominated era. Since we are considering the 4-d projection of the 5-d Schwarzschild metric, such a black hole will have an altered mass-radius relationship given by[30]

$$r_{BH} = \left(\frac{8}{3\pi}\right)^{1/2} \left(\frac{l}{l_4}\right)^{1/2} \left(\frac{M}{M_4}\right)^{1/2} l_4 \tag{28}$$

The black hole will accrete the surrounding radiation with a rate proportional to the surface area of the black hole times the energy density of radiation. The Hawking evaporation rate, as for the case of standard black holes, is proportional to the surface area times the fourth power of temperature[37]. The Hawking temperature is given by

$$T_{BH} = \frac{1}{2\pi r_{BH}} \tag{29}$$

Taking into account these effects of accretion and evaporation together, the rate of change of mass $\dot{M}$ of a braneworld black hole is given by

$$\dot{M} = 4\pi r_{BH}^2 \Big(-g_{brane} \sigma T_{BH}^4 + f \rho_R \Big) \tag{30}$$

where g_{brane} is effective number of particles that can be emitted by the black hole (we assume that the black holes can emit massless particles only and take $g_{brane} = 7.25$[30]), f is the accretion efficiency ($0 \leq f \leq 1$)[38], and σ is the Stefan-Boltzmann constant. The black hole also evaporates into the bulk, with a rate proportional to $4\pi r_{BH}^2 g_{bulk} T_{BH}^5$. However, this term is subdominant even for very small black holes[30], and has negligible effect on their lifetimes.

Substituting the expressions for the black hole radius (Eq.(28)), the temperature-radius relation (29), and the energy density of radiation (24), the black hole rate equation (30) in the radiation dominated high energy braneworld era can be written as[39]

$$\dot{M} = -\frac{AM_4^2}{Mt_c} + \frac{BM}{t} \tag{31}$$

where A and B are dimensionless numbers given by

$$A \simeq \frac{3}{(16)^3\pi} \tag{32}$$

$$B \simeq \frac{2f}{\pi} \tag{33}$$

The exact solution for the black hole rate equation is given by[39]

$$M(t) = \left[\left(M_0^2 - \frac{2AM_4^2}{2B-1} \frac{t_0}{t_c} \right) \left(\frac{t}{t_0} \right)^{2B} + \frac{2AM_4^2}{2B-1} \frac{t}{t_c} \right]^{1/2} \tag{34}$$

with M_0 being the formation mass of the black hole at time $t = t_0$. If the black hole is formed out of the collapse of horizon or sub-horizon sized density perturbations, the formation time and mass are related by[30]

$$\frac{t_0}{t_4} \simeq \frac{1}{4}\left(\frac{M_0}{M_4}\right)^{1/2}\left(\frac{l}{l_4}\right)^{1/2} \tag{35}$$

It was argued in Ref.[39] that a black hole so formed continues to grow in size by the accretion of radiation during the high energy radiation dominated era, with its mass given by[38, 39]

$$\frac{M(t)}{M_0} \simeq \left(\frac{t}{t_0}\right)^{B} \tag{36}$$

One needs to next consider, as in the case of primordial black holes in standard cosmology, a certain number density of primordial black holes exchanging energy with the surrounding radiation by accretion and evaporation. As done in Section 2, we will again use the simplifying assumption that all black holes are formed with an average initial mass M_0. We thus assume that at a time t_0 the fraction of the total energy in black holes is β, and the number density of black holes is $n_{BH}(t_0)$. Hence, one has to now solve the coupled cosmological equations for the radiation density $\rho_R(t)$, the matter density in the black holes $M(t)n_{BH}(t)$, and the scale factor $a(t)$ (the counterpart of the Eqs.(15), (16), and (17) in Section 2) for the braneworld scenario. Notwithstanding the absence of such an analysis in the literature till date, one could still gain certain interesting insights into the cosmological evolution, by making a few simplifying assumptions[38, 39], as we now elaborate.

The number density of black holes $n_{BH}(t)$ scales as $a(t)^{-3}$, and thus for a radiation dominated evolution on the brane, one gets

$$(n_{BH}(t)/n_{BH}(t_0)) = (t_0/t)^{3/4} \tag{37}$$

since $a(t) \propto t^{1/4}$. The net energy in black holes grows since accretion dominates over evaporation. The condition for the universe to remain radiation dominated (i.e., $\rho_{BH}(t) < \rho_R(t)$) at any instant t can be derived to be[39]

$$\beta < \frac{(t_0/t)^{B+1/4}}{1 + (t_0/t)^{B+1/4}} \tag{38}$$

If the value of β exceeds the above bound, there ensues an era of matter (black hole) domination in the high energy braneworld phase. Such a phase of matter domination should definitely be over by the time of nucleosynthesis for the cosmology to be viable.

We shall first explore the situation when the cosmology stays radiation dominated up to the time when brane effects are important, i.e., $t \leq t_c$. From Eq.(38), this requires

$$\frac{\beta}{1-\beta} < \left(\frac{t_0}{t_c}\right)^{B+1/4} \tag{39}$$

Further, the black holes should remain small enough, i.e., $(M/M_4) < (3\pi/4)(t/t_4)$ for the $5D$ evaporation law to be valid[30]. These criteria can be used to put an upper bound to the average initial mass[39]

$$\frac{M_0}{M_4} < \left(\frac{3\pi}{4(2\sqrt{2})^B}\right)^{\frac{2}{2-B}} \frac{t_c}{t_4} \tag{40}$$

The growth of the black holes in the radiation dominated era due to accretion slows down with time, since the surrounding radiation density gets diluted. The rate of evaporation is also insignificant for a wide range of M_0 at this stage since the black hole masses could have grown by several orders of magnitude from their initial values. It is expected that there ensues an era during which the black hole mass stays nearly constant over a period of time, as is the case for standard cosmology (Section 2). The accretion rate is smaller for the braneworld case since the surface area is $\propto M$ instead of M^2 for $4D$ black holes. Moreover, the evaporation rate is ($\propto M^{-1}$) instead of M^{-2}. Hence, the black hole mass will stay for while near a maximum value M_{max} reached at time t_t before evaporation starts dominating. The expression for the lifetime t_{end} of a black hole in this scenario is given by[39]

$$\frac{t_{end}}{t_4} \simeq \frac{4}{A}(2\sqrt{2})^B \left(\frac{M_0}{M_4}\right)^{2-B} \frac{t_c}{t_4} \left(\frac{t_t^2}{t_c t_4}\right)^B \tag{41}$$

It is important to note that the modified evaporation law also contributes to the increased lifetime for braneworld black holes[30]. However, the effect of accretion is more significant, as can be seen by comparing the lifetime of a 5-d black hole in the presence of accretion to the lifetime due to purely an altered geometry[39]. Depending upon the values taken by the parameters t_c and l, one could obtain several interesting examples of primordial black holes surviving till various cosmologically interesting eras[38, 39]. One particular choice worth mentioning is that of a black hole formed with an initial mass $M_0 = 10^8 M_4 \simeq 10^3$g, that will survive up to the present era if one chooses $(l/l_4) \simeq 10^{30}$.

Let us now explore the case of matter (black hole) domination in the high energy phase. The onset of such a matter dominated era is derived to be[39]

$$\frac{t_{heq}}{t_0} = \left(\frac{1-\beta}{\beta}\right)^{\frac{4}{4B+1}} \equiv \gamma_B \tag{42}$$

The mass of a black hole at t_{heq} is given by $M(t_{heq}/M_0) = \gamma_B^B$. For $t > t_{heq}$ the Hubble expansion is essentially driven by the black holes ($p = 0$) which dominate over radiation. Since the number density of black holes scales as matter ($n_{BH}(t) \propto a^{-3}$), for $t < t_c$ one has $H \propto \rho_{BH}$, and thus the scale factor grows as

$$a(t) \sim t^{1/3} \tag{43}$$

During this era, the radiation density ρ_R is governed by the equation

$$\frac{d}{dt}\left(\rho_R(t)a^4(t)\right) = -\dot{M}(t)n_{BH}(t)a(t) \tag{44}$$

where the contribution from accretion black holes is comparable to the normal redshifting term ($\rho_R \sim a^{-4}$) because at this stage the black holes dominate the total energy density. Some further analysis leads to the following expression for the radiation density[39]

$$\rho_R(t) \approx \gamma_B^{-1}\rho(t_0)\left(\frac{t_{heq}}{t}\right) - \frac{\beta B}{B+1/3}\gamma_B^{1/4}\rho(t_0)\left(\frac{t_0}{t}\right)^{1-B} \tag{45}$$

The black hole mass grows as

$$M(t) = M_0\gamma_B^B \exp\left[3B + \frac{C}{B}\gamma_B^{B+1/4} - 3B\left(\frac{t_{heq}}{t}\right)^{1/3} - \gamma_B^{1/4}\frac{C}{B}\left(\frac{t}{t_0}\right)^B\right] \tag{46}$$

where $C = \frac{\beta B}{B+1/3}$. As expected, the accretion regime lasts for a brief duration in the matter dominated phase, beyond which the black hole evaporation starts to play a significant role.

The universe gets reheated as $\rho_R(t)$ increases with time. The stage of black hole domination lasts up to a time t_r ($\rho_R(t_r) = n_{BH}(t_r)M(t_r)$). Subsequently, radiation domination takes over once again. One can derive

$$\frac{t_c}{t_r} \approx \frac{3}{2\delta\gamma_B} + \frac{A}{4\gamma_B^{2B}}\left(\frac{M_4}{M_0}\right)^2 \tag{47}$$

with

$$\delta \equiv \left(\frac{B+1/3}{1-\beta}\right)^{\frac{3}{3\beta+1}} \tag{48}$$

A stringent restriction is imposed by demanding that the standard low energy cosmology (for $t > t_c$) should emerge as radiation dominated. By requiring that the era of black hole domination be over before t_c, i.e., $t_r < t_c$, one gets a lower bound on β from Eq.(35), i.e.,

$$\beta \geq \left[\frac{4t_0}{3t_c}(B+1/3)^{\frac{3}{3B+1}}\right]^{B+1/4} \tag{49}$$

The evaporation time of black holes in this scenario has also been calculated[39]. The black hole lifetime is given by

$$\frac{t_{end}}{t_4} \approx \left(\frac{M_0}{M_4}\right)^2 \frac{t_c}{t_4}\gamma_B^{2B} \tag{50}$$

The effect of accretion is less significant in prolonging black hole lifetimes in this case, as borne out by the following example. For instance, taking the value of the AdS radius to be $l/l_4 \simeq 10^{20}$, and $\beta = 10^{-3}$, black holes with $M_0 \simeq 10^{12}$g evaporate during the present era. This is only to be expected, since early matter domination results in larger hubble expansion rate which further restricts the availability of radiation for the black holes to accrete.

We have thus seen that it is possible to have primordial black holes formed in the high energy era of the braneworld scenario to survive up to several cosmologically interesting eras. Accretion of radiation during the radiation dominated era is primarily responsible for the increased longevity of braneworld black holes, though their 5-d geometry also contributes to a slower rate of Hawking evaporation. It is also worth recalling here that primordial black holes in certain models could also accrete the energy of a cosmological scalar field[14]. Such an effect could lead to further growth of these black holes beyond the early radiation dominated era, thus pushing up their lifetimes further. As mentioned repeatedly, the implications for a population of primordial black holes in cosmology are diversely manifold. At any particular era, the surviving black holes would contribute a portion to the total energy density as dark matter in the universe. If the black holes are produced with an initial mass spectrum, then one would have evaporating black holes at different eras. Hawking radiation from these evaporating black holes would on one hand produce all kinds of particles including heavier ones which could lead to baryogenesis, and on the other, could contribute significantly to the background photons, thus diluting the baryon to photon ratio. Observational constraints impacting different cosmological eras could be used to impose restrictions on the initial mass spectrum of braneworld black holes in a manner similar to the primordial black holes in standard cosmology. Clancy et al[40] have shown how standard constraints are modified in the case of braneworld cosmology. There have been further studies on the impact of braneworld primordial black holes on the high energy diffuse photon background and cosmic ray antiprotons[41]. An interesting question is whether primordial braneworld black holes could contribute to a significant fraction of cold dark matter. In the next section we will investigate an aspect related to the existence of primordial braneworld black holes as cold dark matter.

4 A Mechanism for Binary Formation

In this section we discuss a scheme of binary formation for primordial braneworld black holes. We consider sub-horizon sized black holes in the RSII model formed during the high energy phase. We study the interaction of two or more neighbouring black holes mediated by the surrounding radiation in the radiation dominated high energy era. These black holes exchange energy via the processes of evaporation and accretion with the radiation bath, and through it, with each other. The physical distance between two such neighbours increases initially with the Hubble expansion. Forming out of horizon collapse, the initial mass ratio of two such black holes is on average proportional to the square of the ratio of their formation times (From Eq.(35)). By studying the effect of the interaction term it is possible to show that an initial mass difference between two such black holes can never decrease during the radiation dominated era[42]. Such mass differences facilitate the formation of binaries during the standard low energy phase via three-body gravitational interactions. A formation mechanism for primordial black hole binaries in standard cosmology has been investigated by Nakamura et al.[43]. They showed that if equal mass primordial black holes are inhomogeneously distributed in space, three-body gravitational interactions could lead to the formation of binaries. The scheme of binary formation discussed here could become more effective when combined with the scheme based on spatial inhomogeneities in Refs.[43].

Consider the evolution of two neighbouring black holes, BH1 and BH2, having masses M_1 and M_2 respectively, with initial masses $M_{1_0} < M_{2_0}$, or $t_{1_0} < t_{2_0}$. BH2 forms at a physical distance d_0 from BH1. Their physical separation d grows at a rate $\propto t^{1/4}$. We shall assume that $d_0 > l$, and that the gravitational potential

$$\phi = [M/(M_4^2 d)][1 + (2l^2)/(3d^2)] \ll 1 \tag{51}$$

so that exchange of energy between them via gravitational waves can be neglected during the high energy regime. Both of these black holes are immersed in the radiation bath whose density ρ_R falls off as given by Eq.(24). In isolation, the change of mass of each black hole is effected by the net sum of its Hawking evaporation depending on its temperature, and the accretion of radiation depending on the radiation density in the universe as given by Eq.(31). However, due to the presence of BH2, the net rate of change of the mass $\dot{M}_1$ of BH1 gets a further contribution proportional to the product of $\dot{M}_2$ and the solid angle subtended by BH2 on BH1, and vice-versa. In other words, BH1 feels the local difference in radiation density as a result of an accreting (or evaporating) BH2, and similarly for BH2. Thus the evolution equation for each of them can be written as

$$\dot{M}_i = \frac{BM_i}{t} - \frac{AM_4^2}{M_i t_c} - \frac{r_i^2 \dot{M}_j}{4d^2} \tag{52}$$

for $i, j = 1, 2$, where the Schwarzschild radius at the time of formation is given in terms of the formation mass by Eq.(28), and A and B are given by Eqs.(32). The first and second terms on the r.h.s of Eq.(52) represent accretion from the average radiation density, and evaporation respectively (section 3), and the last term arises from the local inhomogeneity in radiation density due to the j-th black hole. Considering both BH1 and BH2 to be in their accreting phases at the formation time t_{0_2} of BH2, it is possible to see form Eqs.(52) that the for BH1 the first term is comparable to the interaction (third) term when $(d/r_2) \sim ([t - t_{1_0}]/[t - t_{2_0}])^{1/2}$. Thus, for a while after its formation, BH2 is able to suppress the growth of BH1. If on the other hand, BH2 forms at a time when BH1 begins to evaporate, BH2 registers enhanced growth due to the locally denser radiation coming from the evaporating BH1.

The effect of two or more interacting black holes on their evolution has been studied in Ref.[42]. It was shown that interaction with the surrounding radiation always leads to disequilibration of masses, viz., an initial small mass difference between two neighbouring black holes always increases in the high energy radiation dominated era. Some initial mass differences are always to be expected since the black holes formed out of density perturbations at different times have masses given by Eq.(35). Further, any realistic spectrum of initial masses will automatically lead to mass differences for two neighboring black holes formed even during the same era. It was shown[42] that their mass difference at the end of the high energy era always exceeds their initial mass difference. This is a key result that facilitates binary formation, as follows. When the universe exits the high energy phase ($t > t_c$), let the masses of BH1 and BH2 be M_1 and M_2 respectively. Consider a spherical region of radius d in which the average matter density (contributed by BH1 and BH2) is given by

$$\bar{\rho}_{BH} = \frac{3(M_1 + M_2)}{4\pi d^3} \tag{53}$$

As radiation density in this region falls off like a^{-4}, the spherical region of radius d will become matter dominated at a time t_f given by (using Eqs.(24), (25) and (53))

$$t_f \simeq 2\left(\frac{8}{3\pi}\right)^3 \left(\frac{d_0}{r_{2_0}}\right)^6 \left(\frac{t_0}{t_c}\right)^{2B+1/2} t_c \tag{54}$$

When $t > t_f$ the matter dominated region with radius d gets cuts-off from the surrounding radiation dominated expansion of the universe which continues up to $t_{eq} \gg t_c$. For $t > t_f$, the black holes BH1 and BH2 will form a bound system. The size of this bound system d_f is related to the initial black hole separation d_0 by[42]

$$d_f = \frac{1}{16}\left(\frac{t_c}{t_0}\right)^{3-B} \left(\frac{d_0}{l}\right)^3 d_0 \tag{55}$$

Nakamura et al[43] formulated a scheme of binary formation based on the 3-body gravitational interaction of primordial black holes which are spatially inhomogeneously distributed in standard cosmology. The scheme that we discuss here is based on mass differences for primordial braneworld black holes. It is worth noting that in any realistic situation there will both be mass and spatial inhomogeneities, and binary formation proceeds as a result of 3-body interaction aided by the combination of both kinds of inhomogeneities. However, to simplify the analysis here we consider the effect of mass differences only and assume a uniform spatial distribution of black holes. Recall that interaction via the surrounding radiation results in mass differences between neighboring black holes. Consider three neighbouring black holes to have mass ratios at t_c given by $M_1 = \delta_{12} M_2$ and $M_2 = \delta_{23} M_3$ with $\delta_{12}, \delta_{23} < 1$. Such a scenario will arise generically from initial mass differences due to horizon sized collapse at different eras, and possible interactions with other black holes. The region encompassing BH1 and BH2 will decouple from the background Hubble expansion at a time that can be computed using Eq.(54), when BH1 becomes gravitationally bound to BH2 with an elliptical orbit whose major axis a_f is

$$a_f \propto d_f \tag{56}$$

BH3 provides the required tidal force for a stable orbit. The minor axis b_f of the binary is proportional to the tidal force times the squared free fall time, which is given by

$$b_f \propto \left(\frac{2\delta_{12}\delta_{23}}{(1+\delta_{12})}\right)\left(\frac{d_f}{d_P}\right)^3 a_f \tag{57}$$

where d_P is the distance of BH3 from the mid-point of the BH1-BH2 axis. Although BH1 binds to BH2 at time t_f, mass differences hamper the possibility of BH3 being gravitationally bound to BH1. In fact, even if all the black hole separations are equal before binary formation, the total region encompassing the binary and BH3 will be matter dominated at a time much later than t_f. BH3 continues to move away from the BH1-BH2 binary with the Hubble flow while d_P registers growth by a factor $[(t_{md})_{13}/t_f]^{1/2}$[42]. The gravitational potential of the binary at BH3 is much weaker compared to the BH1-BH2 interaction. The role of the third black hole BH3 is to simply provide the tidal force necessary to prevent headlong collision of BH1 and BH2.

The eccentricity of the binary orbit can be obtained from Eqs. (56) and (57). For uniformly distributed black holes formed around the era t_0 with masses $\sim M_0$, the eccentricity is given by

$$\bar{e} = \left[1 - \eta^2\left(\frac{2\delta_{12}\delta_{23}}{1+\delta_{12}}\right)^2\right]^{1/2} \tag{58}$$

where η is a geometrical factor of order 1[43]. The Newtonian approximation that we have used for describing the gravitational interaction between black holes forming a binary is valid if the gravitational potential for the BH1-BH2 binary is such that

$$(M_1/M_4^2) \ll d_f(1-e) \tag{59}$$

From Eqs.(55) and (58) it is possible to show that the Newtonian analysis is suitable for black holes which are formed at times greater than a certain formation time t_0 given by

$$(t_4/t_0)^{2B-1} \ll \eta^2 \alpha_{m0}^{-4/3}(\delta_{12}\delta_{23}/1+\delta_{12})^2 \tag{60}$$

where $\alpha_{m0} = \rho_{BH}(t_0)/\rho_R(t_0)$.

Present interest in the study of binaries stems from the emission of gravitational radiation during their coalescing stage. It is expected that future gravitational wave detectors such as the EURO detector would be sensitive to gravitational waves coming from binaries in the sub-lunar range[24]. The above mechanism for binary formation is suitable for the formation of such binaries within a reasonable range of parameters, as was shown in Ref.[42]. In addition, the scheme based on inhomogeneous spatial distribution[43] of black holes could also act in conjunction to aid binary formation. The range of parameters could be further widened by considering accretion of radiation in the standard radiation dominated phase (section 2) and/or accretion of scalar field dark energy[11, 14]. This calls for further work in this area including accurate determinations of the coalescence time for various models and parameters.

5 Conclusions

The field of primordial black holes in cosmology is a fascinating subject which has engaged several cosmologists over the last thirty years or so. Black holes could be formed in the universe through several mechanisms[1] and in a wide mass range. Their impact on cosmological evolution could be diverse, and is yet to be investigated in full details for different models of black hole formation and cosmological evolution. Two notable consequences of surviving primordial black holes could be the generation of the baryon asymmetry in the universe[8, 9, 15, 10], and their viability as candidates of cold dark matter[20, 21]. In the present article we have studied the cosmological evolution of the early universe with a population of primordial black holes. The effects of Hawking radiation and accretion of energy from the surrounding radiation counteract on the black holes, and thereby impact the whole cosmological evolution. We have carried out this exercise in the context of standard cosmology and also in the braneworld scenario separately. Their are key differences between the above two scenarios.

In standard cosmology the effectiveness of accretion in the growth of black hole mass is restricted because of the fact that the cosmological Hubble radius grows at the same rate as the individual black holes. The role played by accretion here is essentially to counterbalance Hawking evaporation. Therefore, although the net mass increase for a black hole is not quite significant, its lifetime could undergo notable increase[8, 9]. In the braneworld scenario, the first difference stems from the modified geometry of 5-dimensional black holes. This leads to a lower evaporation rate thus increasing the lifetime[30]. The braneworld cosmological evolution is also rather different from the standard one in the early high energy phase[26, 31]. In the radiation dominated high energy phase the rate of mass change for a black hole (Eq.31) can be integrated to give mass growth at a rate lower than the rate of increase of the Hubble radius. Accretion is clearly the dominant effect in the early radiation dominated high energy phase[38, 39]. This could lead to significant rise in the lifetime for primordial black holes with multifarious consequences.

One such consequence that we have discussed in this article is that of binary formation by primordial black holes. If black holes form a significant part of cold dark matter, then a large number of them are expected to be in the form of binaries. Gravitational interaction between neighbouring primordial black holes could start the formation of binaries at early cosmological eras[43]. Braneworld evolution amplifies initial mass differences between neighboring black holes[42], and this effect favors binary formation. Binaries emit gravitational waves during their coalescing stage, and therefore are of observational relevance. So there exists an exciting possibility of obtaining the signature of the physics of primordial black holes in the early universe via gravitational waves from binaries[24]. Of course, if primordial black holes are present in galactic haloes, they should be amenable to detection by gravitational lensing experiments. However, for braneworld black holes the standard lensing results are again modified[44]. This calls for detailed reanalysis of the lensing results pertaining to dark compact halo objects[25].

Cosmological evolution with primordial black holes contains several rich features. One possibility is that of early black hole domination of the energy density whereby the rate of hubble expansion is enhanced. Subsequently, evaporation of the black holes could lead to reheating in the universe affecting several cosmological processes. The magnitude and epoch of such reheating depends on the initial masses and number density of the black holes, since black holes with different initial masses evaporate out at different times. Observational results pertaining to different cosmological eras can therefore be used to impose constraints on the formation parameters of the black holes[19]. Though for reasons of simplicity we have performed the above analysis of the cosmological evolution of black holes with uniform initial masses, formation mechanisms generally give rise to an initial mass spectrum[2, 3, 5]. Due to an altered cosmological evolution, the observational constraints on the initial distribution of black holes are substantially modified in the braneworld scenario[40]. Accretion of radiation affects these constraints since the rate of accretion is higher for bigger initial masses in both the standard and braneword scenario. More general accretion like that from a background scalar field[11, 14] is likely to further modify such constraints on the initial mass distribution. A comprehensive analysis of various constraints in the evolution with accretion, though an arduous task for arbitrary distribution functions, needs to be carried out to shed more light on the different aspects of cosmology with primordial black holes.

References

[1] For a review, see, B. J. Carr, in *The Renaissance of General Relativity*, eds. G.Ellis et al. (Cambridge Univ. Press, 1993).

[2] B. J. Carr and J. E. Lidsey, *Phys. Rev. D* **48**, 543 (1993); A. M. Green and A. R. Liddle, *Phys. Rev. D***56**, 6166 (1997).

[3] K. Jedamzik and J. C. Niemeyer, *Phys. Rev. D***59** 124014 (1999); A. M. Green and A. R. Liddle, *Phys. Rev. D***60**, 063509 (1999); G. D. Kribs, A. K. Leibovich and I.Z. Rothstein, *Phys. Rev. D***60**, 103510 (1999).

[4] For a recent review, see, D. Scott and G. Smoot, *Phys. Lett. B***592**, 1 (2004).

[5] A. A. Starobinsky, *JETP Lett* **55**, 489 (1992); T. Bringmann, C. Keifer and D. Polarski, *Phys. Rev. D***65**, 024008 (2002); D. Blais, T. Bringmann, C. Keifer and D. Polarski, *Phys. Rev. D***67**, 024024 (2003).

[6] Ya. B. Zel'dovich and I. D. Novikov, *Astron. Zh.* **43**, 758 (1966); S. W. Hawking, *Commun. Math. Phys.* **43**, 199 (1975).

[7] B. J. Carr and S. W. Hawking, *MNRAS* **168**, 399 (1974); S. Hacyan, *Astrophys. J.* **229**, 42 (1979).

[8] A. S. Majumdar, P. Das Gupta and R. P. Saxena, *Int. J. Mod. Phys. D***4**, 517 (1995).

[9] P. S. Custodio and J. E. Horvath, *Phys. Rev. D***58**, 023504 (1998); *Phys. Rev. D***60**, 083002 (1999).

[10] N. Upadhyay, P. Das Gupta and R. P. Saxena, *Phys. Rev. D***60**, 063513 (1999).

[11] T. Jacobson, *Phys. Rev. Lett.* **83**, 2699 (1999); T. Harada and B. J. Carr, astro-ph/0412135.

[12] For a review, see, B. Ratra and P. J. E. Peebels, *Rev. Mod. Phys.* **75**, 559 (2003).

[13] D. Richstone et al., *Nature* **395**, A14 (1998).

[14] R. Bean and J. Magueijo, *Phys. Rev. D***66**, 063505 (2002).

[15] J. D. Barrow, *MNRAS* **192**, 427 (1980); D. Lindley, *MNRAS* **198**, 317 (1981); J. D. Barrow, E. J. Copeland, E. W. Kolb and A. R. Liddle, *Phys. Rev. D* **43**, 984 (1991).

[16] V. A. Kuzmin, V. A. Rubakov and M. E. Shaposhnikov, *Phys. Lett. B***155**, 36 (1985).

[17] Ya. B. Zeldovich, A. A. Starobinsky, M. Yu Khlopov and V. M. Chechetkin, *Sov. Astron. Lett.* **3**, 3 (1977); S. Miyama and K. Sato, *Prog. Theor. Phys.* **59**, 1012 (1978).

[18] D. N. Page and S. W. Hawking, *Astrophys. J* **206**, 1 (1976); G. F. Chapline, *Nature* **256**, 251 (1975); J. H. MacGibbon and B. J. Carr, *Astrophys. J.* **371**, 447 (1991).

[19] B. J. Carr in *Observational and Theoretical Aspects of Relativistic Astrophysics and Cosmology*, eds. J. N. Sanz and L. J. Goecoechia (World Scientific, 1985).

[20] D. Blais, C. Keifer and D. Polarski, *Phys. Lett. B***535**, 11 (2002).

[21] N. Afshordi, P. McDonald and D. N. Spergel, *Astrophys. J.* **594**, L71 (2003).

[22] See for example, EROS collaboration, *Astron. Astrophys.* **400**, 951 (2003); P. Popowski et al., astro-ph/0304464; J. Ziolkowski, astro-ph/0307307; P. A. Charles and M. J. Coe, astro-ph/0308020.

[23] Q. Yu, *MNRAS* **331**, 935 (2002); M. Milosavljevic and D. Marritt, *Astrophys. J.* **596** (2003); Q. Yu and S. Tremaine, astro-ph/0309084.

[24] K. T. Inoue and T. Tanaka, P*hys. Rev. Lett.* **91**, 021101 (2003).

[25] *C. Alcock et al., Astrophys. J.* **499**, L9 (1998); G. F. Marani et al., Astrophys. J. **512**, L13 (1999).

[26] R. Maartens, [gr-qc/0312059] *Living Rev. Rel.* **7**, 1 (2004).

[27] L. Randall and R. Sundrum, *Phys. Rev. Lett.* **83**, 4690 (1999).

[28] J. C. Long et al., *Nature* **421**, 922 (2003).

[29] P. Argyres, S. Dimopoulos and J. March-Russel, *Phys. Lett. B***441**, 96 (1998); N. Dadhich, R. Maartens, P. Papadopoulos and V. Rezania, *Phys. Lett. B***487**, 1 (2000); R. Emparan, G. T. Horowitz, and R. C. Myers, *JHEP* **0001**, 007 (2000); R. Casadio and B. Harms, *Phys. Rev. D***64**, 024016 (2001); S. Hemming and E. Keski-Vakkuri, *Phys. Rev. D***64** 044006 (2001); V. P. Frolov, D. V. Fursaev and D. Stojkovic, *JHEP* **0406**, 057 (2004).

[30] R. Guedens, D. Clancy and A. R. Liddle, *Phys. Rev. D***66**, 043513 (2002).

[31] P. Brax, C. Bruck, A-C. Davis, *Rept. Prog. Phys.* **67**, 2183 (2004).

[32] For a review, see J. E. Lidsey, [astro-ph/0305528] *Lect. Notes Phys.* **646**, 357 (2004).

[33] P. Binetruy, C. Deffayet, U. Ellwanger, and D. Langlois, *Phys. Lett. B***477**, 288 (2000); J. D. Barrow and R. Maartens, *Phys. Lett. B***532**, 153 (2002).

[34] R. C. Myers and M. J. Perry, *Ann. Phys.* **172**, 304 (1986).

[35] T. Tanaka, *Prog. Theor. Phys. Suppl.* **148**, 307 (2003); R. Emparan, A. Fabbri and N. Kaloper, *JHEP* **0208**, 043 (2002); R. Emparan, J. Garcia-Bellido and N. Kaloper, *JHEP* **0301**, 079 (2003).

[36] D. M. Eardley and S. B. Giddings, *Phys. Rev. D***66**, 044011 (2002); S. B. Giddings, *Gen. Rel. Grav.* **34**, 1775 (2002).

[37] For a review of braneworld black hole evaporation, see, P. Kanti, *Int. J. Mod. Phys. A***19**, 4899 (2004).

[38] D. Clancy, R. Guedens and A. R. Liddle, *Phys. Rev. D***66**, 083509 (2002).

[39] A. S. Majumdar, *Phys. Rev. Lett* **90**, 031303 (2003).

[40] D. Clancy, R. Guedens and A. R. Liddle, *Phys. Rev. D***68**, 023507 (2003).

[41] Y. Sendouda, S. Nagataki and K. Sato, *Phys. Rev D***68**, 103510 (2003); Y. Sendouda, K. Kohri, S. Nagataki and K. Sato, astro-ph/0408369.

[42] A. S. Majumdar, A. Mehta and J-M. Luck, astro-ph/0311148, To appear in *Phys. Lett. B.*

[43] T. Nakamura, M. Sasaki, T. Tanaka and K. S. Thorne, *Astrophys. J.***487**, L139 (1997); K. Ioka, T. Chiba, T. Tanaka and T. Nakamura, *Phys. Rev. D***58**, 063003 (1998).

[44] A. S. Majumdar and N. Mukherjee, astro-ph/0403405; E. F. Eiroa, gr-qc/0410128; R. Whisker, astro-ph/0411786.

INDEX

G

H

I

K

L

M

N

O

P

Q

R

S

T

U

V

W